Design Innovative Robots with LEGO SPIKE Prime

Seven creative STEM robotic designs to challenge your mind

Aaron Maurer

BIRMINGHAM—MUMBAI

Design Innovative Robots with LEGO SPIKE Prime

Group Product Manager: Rahul Nair

Publishing Product Manager: Vijin Boricha

Senior Editor: Arun Nadar

Content Development Editor: Sujata Tripathi

Technical Editor: Shruthi Shetty

Copy Editor: Safis Editing

Project Coordinator: Shagun Saini

Proofreader: Safis Editing

Indexer: Manju Arasan

Production Designer: Joshua Misquitta

First published: January 2022
Production reference: 2071221

Published by Packt Publishing Ltd.
Livery Place
35 Livery Street
Birmingham
B3 2PB, UK.

ISBN 978-1-80181-157-6

www.packt.com

To my wife, who constantly challenges me to become a better person and for pushing me to dig within myself to believe in what I have to offer to the world.

To my three children, Aiden, Addy, and Ava, who remind me to hold onto my crayons to remember the joys of being a child and living in and embracing the moment. As adults, it is easy to lose sight of the power of play and enjoying the moments.

To all the people who have supported and followed me in my journey. Continue to make the world a better place one brick at a time.

Note from the Author

Welcome to this book. I am so excited to have the opportunity to share with you a passion of mine as an educator who has hands-on experience with LEGO, transforming learning and engagement with students; a parent who has spent countless hours building and creating stories with my kids; an **Adult Fan of LEGO** (**AFOL**) with my own LEGO collection to fulfill my hobby; and, of course, just someone who loves the energy of the LEGO community and the continuous positive sharing of remarkable ideas.

I have one major goal with this book: I want it to be a permission slip for you to build and make. Sometimes when we read books such as these, we build as suggested in the book and that is the extent of the creative process. Being an educator who has spent time in elementary and middle schools, I have learned that not providing the answer is the true facilitator of creativity. What I have created is a series of builds for you to explore not only the LEGO SPIKE Prime Kit but also the world of smart robots that impact our daily lives.

So, each chapter will explore a smart robot concept along with a model robot to build. However, each build has been designed for you to add your own touch. You will see suggestions and ideas to apply the learning of the chapter to new layers of building. I hope you love the freedom to take the ideas and build your own version. With this permission slip, I cannot wait to see what you create. Make sure you share! Here we go. Let's go explore the new LEGO SPIKE Prime Kit that is now available.

Contributors

About the author

Aaron Maurer, also known as *Coffeechug* is the STEM lead for 21 school districts in Iowa helping to expand STEM, Computer Science, Makerspace, and Purposeful Play into classrooms K-12. Aaron was also a FIRST LEGO League coach for 8 years with much success working with phenomenal kids.

He has a Master Educator License with endorsements in 5-12 Psychology – 163; 5-12 World History – 166; 5-12 American History – 158; 5-12 Computer Science – 278; PK-12 Talented and Gifted – 107; 5-8 Middle School Generalist – 182; and K-8 Computer Science – 277.

Currently, Aaron is a member of ISTE Making It Happen Award; LEGO Education Ambassador and Master Educator; PBS Digital Innovator and All-Star; PITSCO Tag Committee; Microsoft Innovator Educator Expert; Microsoft Innovative Educator Fellow; Global Learning Mentor (formerly Skype Master Teacher); Minecraft Global Mentor; and Makey Makey Ambassador.

Finally, Aaron was a finalist for the Iowa Teacher of the Year in 2014.

You can find more of his work at www.coffeeforthebrain.com.

About the reviewers

Khushboo Samkaria (Rana) lives in India with her husband, Vivek, and son, Agastya. She is a certified **LEGO Serious Play** (**LSP**) practitioner with more than 8 years of experience working globally with LEGO Education partners and organizing **FIRST** (**For Inspiration and Recognition of Science and Technology**) programs. She is devoted to the mission of helping young innovators to understand the real world using an interdisciplinary approach – **STEM** (**Science, Technology, Engineering, Math**). Since 2011, she has interacted with more than 100,000 kids and adults globally as a part of the FIRST programs. Currently, she is working with various organizations to help them tap their knowledge and ideas using the LSP approach so that they can be the winner in this period of intense change.

Scott Reece holds an Educational Specialist in mathematics with a certification in technology education. His career as a classroom teacher spans 26 years in social studies, mathematics, and technology classrooms. Even before becoming an educator, Scott was working with children through the children's ministry at different churches. His passion is seeing children grow, both spiritually and educationally. Over the years, Scott has become a *go-to* for technical support both in his school and in his district. Scott also began a relationship with LEGO Education that led to him being recognized as a US Master Educator in 2019. Scott also proudly holds the roles of husband, father, and DeDe (grandpa).

Table of Contents

3

Building a LEGO Guitar

4

Building a Mechanical Bird

5

Building a Sumobot

6

Building a Dragster

7

Building a Simon Says Game

Other Books You May Enjoy

Index

Preface

The new LEGO SPIKE Prime is one of the latest additions to the LEGO robotics line of products. This book will help you to enjoy building robots and understand how exciting robotics can be in terms of design, coding, and the expression of ideas.

The book begins by taking you through a new realm of playful learning experiences designed for inventors and creators of any age. In each chapter, you'll find out how to build a creative robot, learn to bring the robot to life through code, and finally work with exercises to test what you've learned and remix the robot to suit your own unique style. Throughout the chapters, you'll build exciting new smart robots such as a handheld game, a robotic arm with a joystick, a guitar, a flying bird, a sumobot, a dragster, and a Simon Says game.

By the end of this LEGO book, you'll have gained the knowledge and skills you need to build any robot that you can imagine.

Who this book is for

This book is for robot enthusiasts, LEGO lovers, hobbyists, educators, students, and anyone looking to learn about the new LEGO SPIKE Prime kit. The book is designed to go beyond the basic builds to intermediate and advanced builds, while also helping you to learn how to add your own personal touch to the builds and code. To make the most of this book, you'll need a basic understanding of build techniques, coding in block-based software environments, and weaving them together to create unique robot builds.

What this book covers

Chapter 1, Getting Started with SPIKE Prime, focuses on the new kit and explores all the elements, sensors, components, and the coding platform that come with the kit. It is important to understand the new pieces introduced in this kit along with the sensors and how they work.

Chapter 2, Building an Industrial Robot Claw, contains our first robotic project, in which we explore how smart robots are used in everyday life. We will explore the concept of a claw and how it is used in various jobs and industries. Finally, we will build a claw that can be expanded by the user to complete various build challenges.

Chapter 3, Building a LEGO Guitar, explains that entertainment is important to our well-being. This chapter will explore how to build a working guitar that can be played and modified to meet the needs of the user.

Chapter 4, Building a Mechanical Bird, explores the mechanisms required to design and build a mechanical bird. It is designed to showcase new ways to build robots using mechanisms for wing, head, and leg movement.

Chapter 5, Building a Sumobot, covers the sumobot robotic challenge, a classic robot challenge that has been around for a long time. Despite not being a new concept, it is loved by many. This chapter contains a sumobot build to take your competitive advantage to new levels.

Chapter 6, Building a Dragster, focuses on designing and building a dragster to see how fast we can move our robot down a speedway. Robots have been playing a critical role in travel.

Chapter 7, Building a Simon Says Game, contains a very fun project for hobbyists to build and code: a handheld game of Simon Says. This chapter will explore less on the actual build mechanics; instead, more space is given to the developing code. The goal of this chapter is to increase your coding skills so you can go and create a game of your own.

To get the most out of this book

You require a basic understanding of build techniques, coding in block-based software environments, and how to weave them together to create unique robot builds.

Software/hardware covered in the book	OS requirements
The LEGO SPIKE Prime software	Windows, macOS, iOS, Android, Fire OS
LEGO SPIKE Prime Kit 45678	

If you are using the digital version of this book, we advise you to type the code yourself or access the code via the GitHub repository (link available in the next section). Doing so will help you avoid any potential errors related to the copying and pasting of code. The programming software can be used on computers and/or tablets/phones. You may want to explore and experiment with both platforms to find out what you prefer.

Download the example code files

You can find the code files for this book on GitHub at: `https://github.com/PacktPublishing/Design-Innovative-Robots-with-LEGO-SPIKE-Prime`. If there's an update to the code, it will be updated in the GitHub repository.

We have other code bundles from our rich catalog of books and videos available at `https://github.com/PacktPublishing/`. Check them out!

Code in Action

Code in Action videos for this book can be viewed at `https://bit.ly/3r0qpSy`.

Download the color images

We also provide a PDF file that has color images of the screenshots/diagrams used in this book. You can download it here: `https://static.packt-cdn.com/downloads/9781801811576_ColorImages.pdf`.

Conventions used

There are a number of text conventions used throughout this book.

`Code in text`: Indicates code words in text, database table names, folder names, filenames, file extensions, pathnames, dummy URLs, user input, and Twitter handles. Here is an example: " …and select numbers `1-3`."

Bold: Indicates a new term, an important word, or words that you see onscreen. For example, words in menus or dialog boxes appear in the text like this. Here is an example: "Add an orange control block named **stop all**."

> **Tips or important notes**
> Appear like this.

Get in touch

Feedback from our readers is always welcome.

General feedback: If you have questions about any aspect of this book, mention the book title in the subject of your message and email us at `customercare@packtpub.com`.

Errata: Although we have taken every care to ensure the accuracy of our content, mistakes do happen. If you have found a mistake in this book, we would be grateful if you would report this to us. Please visit www.packtpub.com/support/errata, selecting your book, clicking on the Errata Submission Form link, and entering the details.

Piracy: If you come across any illegal copies of our works in any form on the Internet, we would be grateful if you would provide us with the location address or website name. Please contact us at copyright@packt.com with a link to the material.

If you are interested in becoming an author: If there is a topic that you have expertise in and you are interested in either writing or contributing to a book, please visit authors.packtpub.com.

Share your thoughts

Once you've read *Design Innovative Robots with LEGO SPIKE Prime*, we'd love to hear your thoughts! Please click here to go straight to the Amazon review page for this book and share your feedback.

Your review is important to us and the tech community and will help us make sure we're delivering excellent quality content.

1
Getting Started with SPIKE Prime

In this chapter, we will be exploring the **SPIKE Prime** kit. It has over 500 elements to build robots, upgraded sensors, a new Intelligent Hub compared to previous versions of MINDSTORMS, and new programming platforms to bring our robots to life and control them.

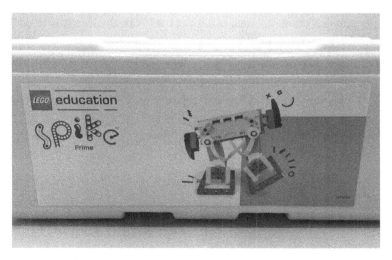

Figure 1.1 – The front of the box when you get your hands on this amazing kit

To begin with, we are going to examine what the new elements are in this kit for us to build our designs with, how the sensors have changed, and the new Intelligent Hub. As we explore, we will take a closer look at some excellent features of this kit to start learning more about this LEGO MINDSTORMS product.

Finally, we will do some exploration of the new programming interface and how we use code to bring our ideas to life. The end goal of this chapter is to make sure you understand all possibilities and to build a foundation of basic knowledge of the kit to begin to build some of the exciting ideas to follow in the upcoming chapters. This would also be a perfect time to explore the **Getting Started** section of the software to see how everything works and operates if you are new to SPIKE Prime. You will build one mini project at the end of this chapter to serve your understanding of what can be achieved with the parts.

In this chapter, we're going to cover the following main topics:

- Overview of the kit
- Intelligent Hub
- New elements
- Sensors
- Programming
- Basic projects to learn more

Technical requirements

One of the creative constraints when designing the builds in this book is to only use the parts, sensors, and elements contained in the kit. There will not be any additional parts needed to complete any of these builds. You will only need the SPIKE Prime Set 45678.

For software, you will need to download the LEGO SPIKE Prime software on either your computer, phone, or tablet. Please check the site to ensure your hardware and operating system are compatible with the software. You can get everything you need to set up by choosing LEGO Education SPIKE Prime on the LEGO site at `https://education.lego.com/en-us/start`.

For the building of the robot, all you will need is the SPIKE Prime kit. For programming, you will need the **LEGO SPIKE** application/software.

Access to the code can be found here: `https://github.com/PacktPublishing/Design-Innovative-Robots-with-LEGO-SPIKE-Prime/blob/main/Ch%20 1%20Rock%20Paper%20Scissor.llsp`.

You can find the code in action video for this chapter here: `https://bit.ly/2Ziax2e`

Overview of the kit

This kit comes with 523 elements to build, design, and bring your ideas to life.

When you open the plastic tub, you'll see two pieces of paper. The first one provides a layout of how LEGO Education suggests you organize the parts and elements in your tub. The second paper contains all the stickers to label and mark the trays, tubs, and elements.

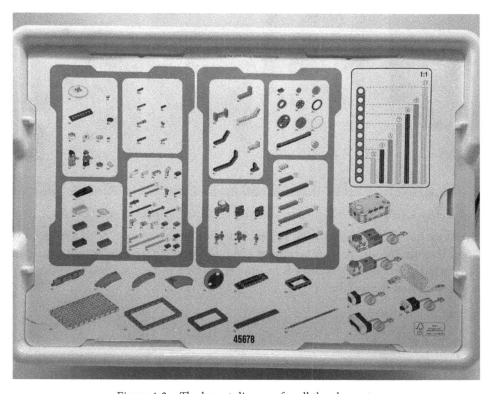

Figure 1.2 – The layout diagram for all the elements

Next, you will find the inside of the box has the bottom of the tub to store the larger pieces and two smaller plastic sorting trays with stickers to better organize and sort your elements.

Figure 1.3 – Two smaller sorting trays to help organize and build more efficiently

The kit does not have a manual to build robots, but the software comes with a huge library collection of lessons, builds, and units to help a builder learn some new build techniques and to get started with building robots. It is quite nice once you get started with the software to see all that it provides.

Here is what the kit provides:

- One Micro USB cable
- One Intelligent Hub
- One rechargeable battery

- Sticker sheet
- One plastic tub with two sorting trays
- Over 500 building elements
- External sensors (one distance sensor, one color sensor, and one force sensor)
- Motors (two medium motors and one large motor)

You will explore these parts in greater detail throughout this chapter to understand them better.

The cables for motors and sensors

One of the biggest changes that the builder will notice right away is the cable connections. If you have previously used LEGO MINDSTORMS kits such as the EV3 or NXT, then you will notice the new Intelligent Hub. Additionally, you will notice that the cables are different. They are flat and smooth. They are also permanently connected to the motors and sensors, unlike previous kits where you could use various lengths of cables to connect and build your robots.

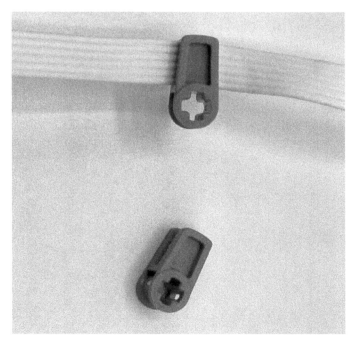

Figure 1.4 – Flat wires and wire clips

The kit comes with clips to help with wire organization and building, which is an excellent new upgrade to the building kits. No more rubber bands and looping cables. These clips are nice to hide cables and manage them to allow your builds to look much more polished (they come in six colors). The downside to these new cables is that these cables, sensors, and motors are no longer compatible with previous kits. If you are like me, then you have a lot of NXT and EV3 parts that you love that are no longer compatible with these new features. Another downside currently is the cables are of fixed length. In previous kits, the cables were separate from the elements, so you had a wide range of wire sizes to fit your build. I am sure it won't be long before third-party companies create extensions, but for the sake of the actual kit, all your wire lengths are the same so plan your builds accordingly.

One thing about this kit that I love is the variety of colors. The kit is designed for middle school students (grades 6-8), but when working with younger students, older students, and adults, there are times when color plays a huge role in how we think and creatively come up with solutions. The wire clips, for example, come in *six* different colors allowing for organizing wires and color-coding to match the build of the robot.

Let's dive into some specific parts and pieces worth exploring.

The Intelligent Hub

Another upgrade to this kit from previous versions is the size of the Intelligent Hub. The Hub is much smaller, which will allow for more unique builds, especially when combined with some of the new elements. The bulkiness of the EV3 and NXT is no longer going to be an issue. The way the Intelligent Hub is designed allows the builder to position the Intelligent Hub in a variety of ways, making building designs easier.

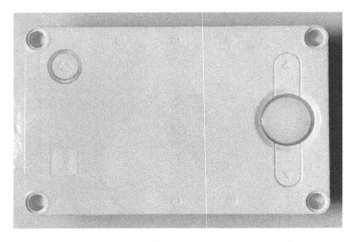

Figure 1.5 – The new Intelligent Hub that comes with the kit

The LED screen is one of the most noticeable changes to the Intelligent Hub. In previous models, we had a screen where the programmer could program on the Intelligent Hub and have various elements of text, data, and graphical images displayed on the screen. On this Intelligent Hub, we are given a 5x5 LED screen. The days of programming on the Intelligent Hub are gone with this new design. However, keep in mind, it is so much easier to get your programs to the Intelligent Hub. Once your Intelligent Hub is paired to your software, you can make changes instantly, so this feature is no longer needed.

The new Intelligent Hub has an LED design that reminds me of my 8-bit glory days of playing Tetris. These LED blocks allow you to gather some basic data on your sensors, switch programs, and write out words that scroll across the screen. Additionally, you use the arrow buttons to scroll through your programs where you have 32 MB of memory available for storing programs, sound, and content.

Using the large button on the Intelligent Hub, you can gather some quick data and test your sensors and motors by plugging them in and gaining some quick information.

The Intelligent Hub has the 6-axis gyro sensor built in. In the EV3, the gyro was a sensor we had to attach much like the touch, color, and ultrasonic sensors. This is a nice touch to the Intelligent Hub, especially now that the Intelligent Hub has been reduced from eight ports to six. While at first you might be bummed to only have six ports, remember this could prove to be helpful as these six ports can be both input and output. Looking back at the EV3, it had four ports for motors and four ports for sensors. The new Intelligent Hub allows the user to use any port for any motor or sensor. If you have been using EV3 for as long as I have, then you'll realize that this means you could potentially have six motors running off the hub or six sensors, where in the previous models we were limited to four or less.

The Intelligent Hub also contains an accelerometer, an internal speaker, and a rechargeable lithium-ion battery. These functions expand the Intelligent Hub to allow the builder to do some exciting projects. We will explore some of these features later, but it is a great reminder that the Intelligent Hub now serves as more than just the brains of the operation. It has several new features to expand what we can do with our robots without having to use ports. For example, you can now run your program tethered to your device or connect via Bluetooth for a wireless experience. Additionally, the Intelligent Hub has an embedded MicroPython operating system to really open up the gateways for possibilities for those that want to take things to the next level of building, design, and code.

Elements

Depending on what previous LEGO MINDSTORMS kits you have used in the past and whether you have any experience with LEGO MINDSTORMS Robot Inventor 51515 or some Powered UP kits, these elements might look familiar.

For the sake of exploring the kit, I would like to highlight a few key elements that I believe are great to have that the previous EV3 kits did not provide.

A quick note about the elements is that while many of the pieces are not necessarily new, there are several that work well together when combined. Additionally, the kit also provides many colors aside from the standard gray, black, and white pieces of previous kits. This will be a huge bonus for many builders looking to expand and coordinate their builds to have a more polished look. I *really* like having choices in how my robots look.

Let's examine a few specific elements that come in the kit that are worth exploring a bit more deeply.

Panel plate

I fell in love with this piece when I purchased the SPIKE Prime kit. The kit provides two of these yellow 11x19x1 plates. The Robot Inventor Kit also provides this piece in teal if looking for another color option.

Figure 1.6 – The 11x19x1 panel plate

This plate allows the builder to build right upon it for stationary robots as well as providing a starting point for other robot builds such as a vehicle or another mechanism. The possibilities this element provides the builder with are huge. It is the piece I wish I always had as a kid.

Wheels

The kit provides four tires, which are great. They are all the same 56 mm size and are all black with teal frictionless wheels. They are one piece, which is nice compared to the previous wheels where the rubber was a separate element from the tire hub. They will be easy to clean for smooth driving, and having four of these wheels to build with will allow the builder to build some excellent designs. The new wheel design is designed to be frictionless to enhance maneuverability and increase precision.

Figure 1.7 – This is the standard tire that comes with the kit

If you are looking for larger wheels that are built like these, there is a pair of larger wheels featured in the SPIKE Prime expansion kit. Remember, you will only need the SPIKE Prime kit for all the builds in this book.

Multiple size frames

There are three sizes of these elements that come in the kit. These frames provide a nice way to build some larger robots and to build more secure structures. You won't realize how much you need these elements until you build with them. Once you have them in your collection, you come to depend on them quite a bit. Again, color is what makes these different, with each size a different color compared to the standard black we typically see.

Figure 1.8 – The three open frame sizes that come with the kit: 11x15, 7x11, 5x7

Now that you have explored a few of the larger key elements, let's dive into some specific parts that are included in the kit.

Integrator brick

The kit comes with what LEGO Education calls new, innovative elements that are featured in this kit for the first time. The integrator brick is your typical 2x4 brick, but in this case, these pieces have three cross axle holes in them. These elements come in five colors. The beauty of these bricks is that they allow combining LEGO Technic and other LEGO platforms together to expand what your imagination comes up with along with all sorts of new building possibilities.

Figure 1.9 – Integrator brick

Some of you will love the gear differential, but not everyone builds with these parts. However, there are some parts that almost everyone will love having in the kit. Let's check some of these out.

Ball and castor

I love that this kit comes with a ball and castor. If you have used LEGO MINDSTORMS EV3, you know how helpful a part like this can be for moving. This new design has a ball, which is plastic instead of metal, and the new castor wheel element design is very useful to quickly design a driveable robot. This is a piece that I felt was missing big time in the LEGO MINDSTORM Robot Inventor Kit 51515.

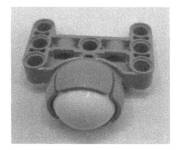

Figure 1.10 – Ball and castor

And while these are great pieces, here is another element in the kit that is even cooler!

Biscuit

This piece might be my favorite due to how useful I repeatedly find this element to be. To the eye, it is nothing special, but this piece allows the builder to strengthen and enhance stability in builds. Additionally, it also allows building options to be multi-directional, which is a huge plus. You have 10 of these items in your kit, 6 being in dark pink and 4 in black.

Figure 1.11 – Biscuit element

Next, let's look at sensors!

Sensors

This kit provides three external sensors. Don't forget that the Intelligent Hub contains a gyroscope sensor, opening ports for new builds compared to previous robot kits. Overall, you are given the following sensors:

Inbuilt sensors:

- 6-axis gyro sensor: 3-axis accelerometer and 3-axis gyroscope
- Speaker
- Gestures: Tap, free fall, and shake

External sensors:

- Color sensor
- Distance sensor
- Force sensor

The Intelligent Hub itself contains an accelerometer and gyroscope, which are great to use the data they collect to write some quality code with our robots. Another neat feature of the Intelligent Hub and these sensors is the gesture controls that allow the builder to create code based on tap, free fall, and shake using these sensors.

The force sensor looks similar in many ways to the touch sensors found in the EV3 and NXT kits from years ago. However, this sensor does more than the touch sensor could do. This sensor allows the user to measure force up to 10 newtons. It also serves as a touch sensor when pressed, released, and bumped.

Figure 1.12 – Force sensor

The color sensor has been upgraded compared to previous versions. The color sensor is able to identify a small dose of color to make decisions. The sensor can also detect eight colors. Finally, it can identify these colors in both dark and bright light, which is very helpful. The sensor allows the coder to use color and reflection light.

Figure 1.13 – Color sensor

Finally, the distance sensor is relatively similar to previous models except for a few changes. First, it has lights around the eye parts of the sensor that can be activated. The builder can program these lights, which is a cool feature.

The sensor is more accurate than previous models, but the range has been reduced from 250 cm to 200 cm. This will not impact many builders but is worth noting. You can choose distance settings of inches, centimeters, or percent.

Figure 1.14 – Distance sensor

Overall, the kit provides the builder with three motors and three sensors along with the sensors built into the Intelligent Hub itself, providing countless opportunities for building and coding.

Motors

This kit provides the builder with three motors. Two of these motors are medium motors and one motor is a large motor. The motors have sensors that allow you to gather data on both speed and position when using the app. One thing you will notice is that these medium motors are smaller than the medium motors of previous kits but are much easier to build with in your designs. This is a nice feature to allow the builder to create more fluid and precise builds. The shape also allows easier builds than previous models where the motors had some unique shapes that could challenge how the builder created their creations.

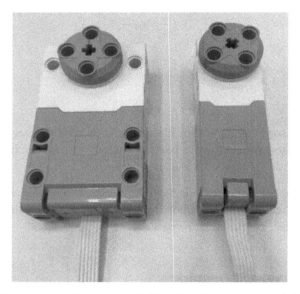

Figure 1.15 – Large motor (left) and medium motor (right)

The new motors are different from the NXT and EV3 motors. In the previous kits, you were able to put an axle all the way through the motor. These new motors do not allow the builder to do this. At first, I did not like this, but I realized quickly that it did not prevent any builds from being successful as you are still able to add axles on both sides of the motor.

One key advantage of these motors is their absolute positioning. This helps with the alignment of the motors and to have more precise positioning when using robots that require motors to be synced.

Here is a quick breakdown of the motors:

	Large motor	Medium motor
Wire length	250 mm	250 mm
No-load speed	175 RPM +/-15%	185 RPM +/-15%
Maximum efficiency torque	8 ncm	3.5 ncm
Maximum efficiency speed	135 RPM +/-15%	135 RPM +/-15%

Table 1.1 – Motor comparison chart

Now that we have a better understanding of the elements of the kit, it is time to explore the software to bring builds to life.

Hub connection

Using the SPIKE Prime software, you can adjust the sensor settings by clicking on the sensor icon and adjusting as needed. The following steps show how you can use the software to gather data on your motors and sensors and make the necessary adjustments for the needs of your build:

1. In order to access the hub connection, you will need to open up a new project or any existing program you have already started. Be sure your Intelligent Hub is also powered on and connected to the software through the cable or Bluetooth.

2. In the upper left-hand corner of the programming canvas, there is an Intelligent Hub icon that you will click on to open the interface.

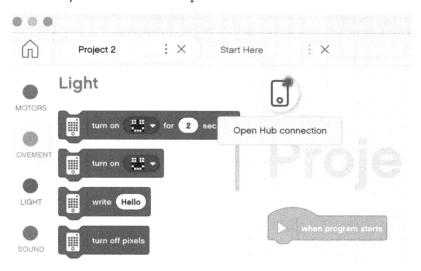

Figure 1.16 – The hub connection icon in the upper-left hand corner.
Green means the hub is connected, and red means disconnected

You will see the following screen, which will showcase all the motors and sensors currently plugged in and activated on the Intelligent Hub.

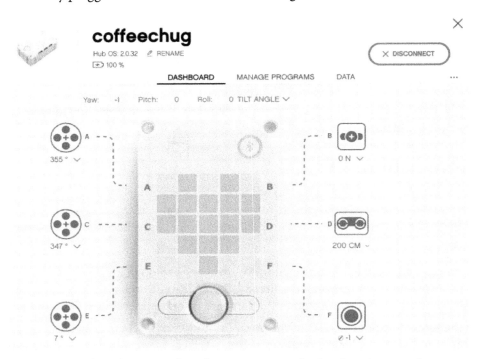

Figure 1.17 – The software interface allowing you to see the data from inputs and outputs

3.   Select a motor or sensor and click the blue arrow below the icon to choose the various types of data readings that you can use for your projects.

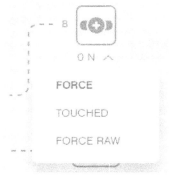

Figure 1.18 – Each motor and sensor will have different options

4. For the inbuilt sensors, click on the sensor you want to see the information for and then move your Intelligent Hub around to see the data readings change.

Figure 1.19 – Change sensor readings just like other inputs/outputs

Now that you understand how to access and connect the Intelligent Hub, it is time to explore the programming options.

Programming

When you boot up the software, you will have the choice to build robots using **Essential** or **SPIKE Prime**. Choose **SPIKE Prime**.

Figure 1.20 – Software provides platforms to choose from

If you need ideas, then head to `https://education.lego.com/en-us/lessons?products=SPIKE%E2%84%A2+Prime+Set` and you will get what you need from this page.

I find having the software on a computer or laptop for coding is quite helpful because of the larger screen and being able to store programs in my system. It is a good idea to have the software installed on your tablet because these are another good option depending on what you have available to use. Explore all the options to find what works best for you as a designer.

To get started, if you are brand new to the kit and LEGO robotics, then I suggest starting with the **START** menu option of the software to give you the option to get a quick overview of the sensors and builds and how they work with the Intelligent Hub.

Getting Started

Figure 1.21 – The software provides Getting Started lessons if you are new

This little build and coding activity is a great start to explore much of the content shared in this chapter.

Once you have built some of these robots and are ready to expand to new ideas, then it is time to begin to build your own projects. When you are ready to create your own builds and programs, you can choose **New Project** from the **Home** screen.

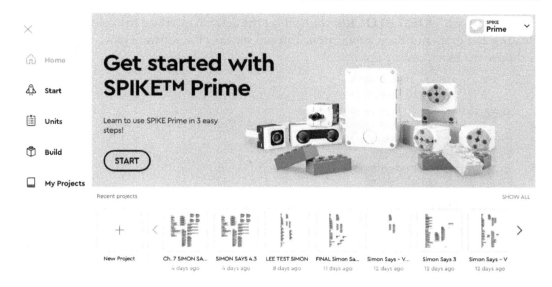

Figure 1.22 – Choose New Project

You will have the choice to program with word blocks or Python as shown in the following screenshot:

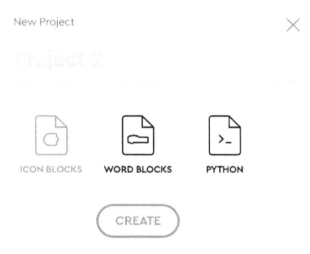

Figure 1.23 – You can choose blocks or Python

When you choose **WORD BLOCKS**, you will get the Scratch interface to explore and build. This is a drag-and-drop application. You can see how the software looks when you choose the block coding option.

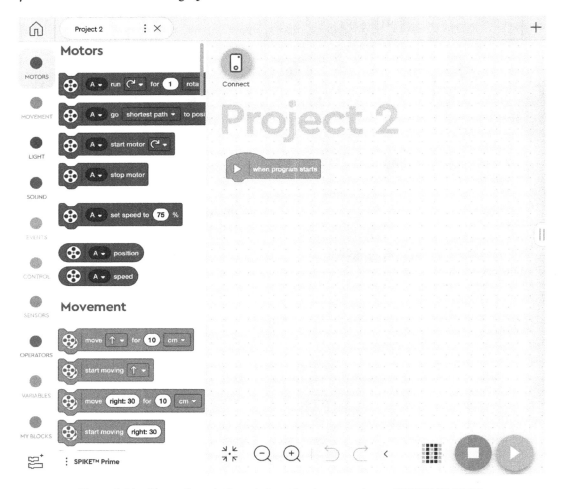

Figure 1.24 – The coding platform is Scratch when you choose WORD BLOCKS

The software is a Scratch-based programming platform, a drag-and-drop coding process that many are already familiar with. You will be quite comfortable with this interface if you have used Scratch, Blockly, MakeCode, or Code.org. There is also a Python coding option within the software, which will allow those who know how to program in Python to really push the boundaries of coding. This is a very important feature to bring this kit to the next level of experience and expertise. The following screenshot showcases what the Python coding interface looks like in the software if you wish to go down this route:

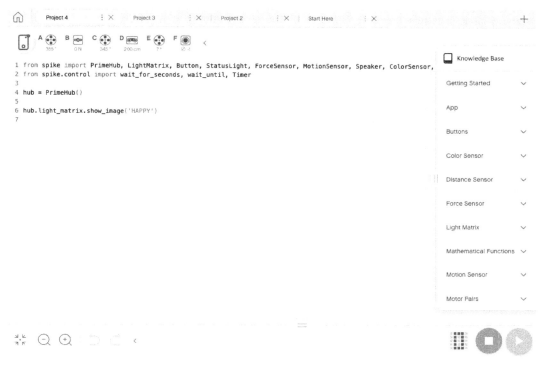

Figure 1.25 – Python coding interface within the software

I know that not everyone knows how to code in Python, so for the sake of this book, we will stick with the graphical interface for ease of understanding.

Having both options to code and program within the software is a huge bonus to prior kits. The option for text programming is going to allow the LEGO community to see some incredible projects developed.

If you are new to Python, there is a **Knowledge Base** option on the right side of the screen to help you see how the blocks look and operate.

Once you click on **Knowledge Base**, you will be provided with a menu of options to help your learning and coding journey. Simply choose the topic you need help with as seen in the following screenshot:

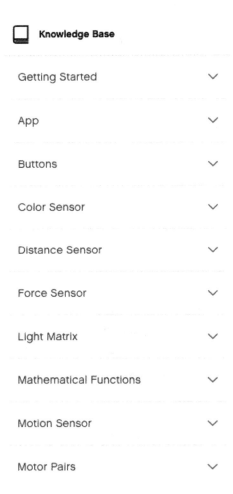

Figure 1.26 – Find the Knowledge Base option on the right side of the screen

As you can see, there is a ton of potential in this kit as you continue to explore and unravel all that can be accomplished.

Let's do one simple build together just to get familiar with building and coding before we dive into the many awesome robot builds to come in the book.

Creating a rock, paper, scissors wrist game

So far, you have been doing a lot of reading, so let's take time to do a quick build that covers some of the concepts shared about this new and exciting kit. You are going to build a fun little game of rock, paper, scissors that you can wear on your wrist.

You will need the following parts:

- One Intelligent Hub
- 16 black connector pins
- One purple *7x11* open frame
- Two black 5x7 open frames
- Two black 9L beams
- Two gray *H* connector brackets
- Eight gray connector pins with a stop bush

Follow these steps to build the rock, paper, scissors wrist game:

1. To start this build, find your Intelligent Hub.

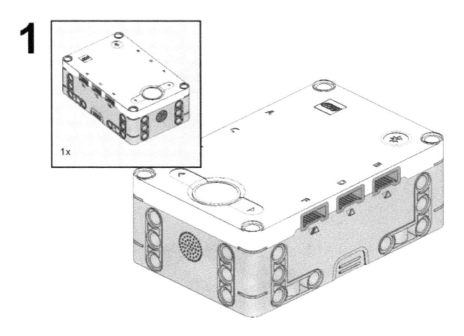

Figure 1.27 – Intelligent Hub

2. Locate four black connector pins. Flip the Intelligent Hub onto its end and insert the four pins into the middle pin holes at the top and bottom of the hub.

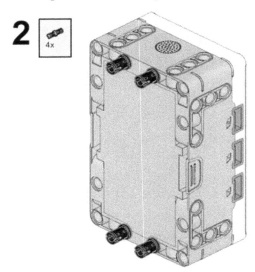

Figure 1.28 – Black pin connectors on the Intelligent Hub

3. Locate your two black *5x7* open frame parts and attach the *5L* side to the pins you just inserted into the Intelligent Hub.

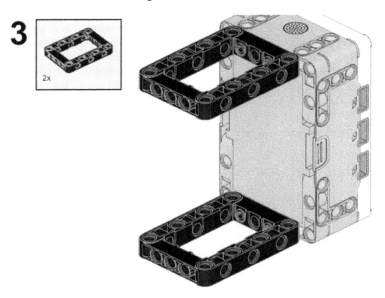

Figure 1.29 – Open frames added to the Intelligent Hub

4. Using two black connector pins with each *9L* beam, secure the open frame and the Intelligent Hub with these beams.

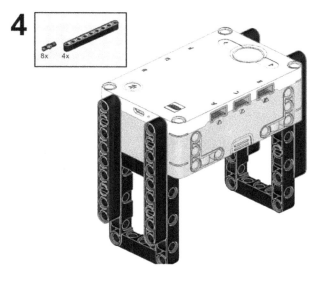

Figure 1.30 – Two 9L beams added to each side

5. Add two black connector pins to the middle pin holes on both sides of the *7L* side of the purple open frame. Connect the purple open frame to the bottom of the black open frames. The purple open frame is the piece you will remove to attach the frame to your wrist before locking everything into place.

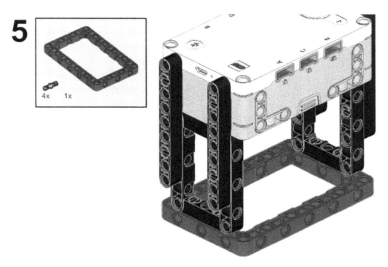

Figure 1.31 – Open frame and black pins

6. Insert four of the gray pins with bush stops into the bottom two holes on each side of the *H* bracket. Do this on both sides.

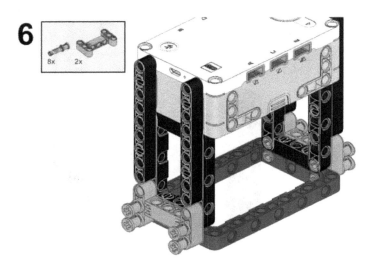

Figure 1.32 – Gray pins and H beams

7. Use the gray pins with bush stops to secure the bottom open frame to the sides of this build. If you want to attach this to your wrist, insert your wrist first and then secure it with these parts. If you have a tiny wrist, then hold onto the side so it does not fall off when you shake it while playing the game!

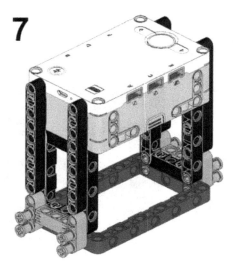

Figure 1.33 – Completed build

This is a very simple build, but it is designed to be quick and easy to assemble to be able to play your first game of rock, paper, scissors. It is now time to write some code and play the game!

Coding a rock, paper, scissors wrist game

You are now going to code this wrist game to be able to randomly choose rock, paper, or scissors. You can play it against yourself or use it against another human. If you have a friend that has a kit, then you can both build one and compete this way. Start by opening up **New Project** and giving your project a name such as **Rock Paper Scissors**:

1. Using the default yellow **when program starts** block, add a purple light block named **write Hello**. You will need to change the word from **Hello** to **Left**.

Figure 1.34 – "Left" being programmed on the screen

2. Add two yellow **Events** blocks named **when Left Button pressed** and **when Right Button pressed**.

3. Go to the pink **My Block** sections of the coding blocks and make two new blocks. Name one **start game** and the other **play again**. Add these new blocks to the yellow blocks. Add the **start game** block for the left button and **play again** for the right button.

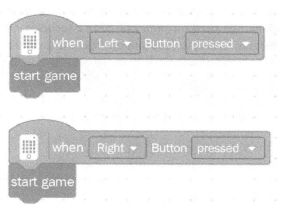

Figure 1.35 – Decision making with My Block

4. In the previous step, when you created these blocks, you should have seen two new pink blocks show up in your coding canvas named **define play again** and **define start game**. Locate the **define play again** block:

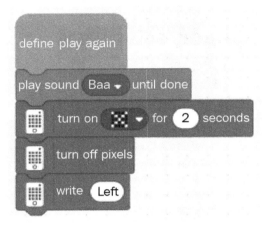

Figure 1.36 – play again My Block

Under the **define play again** block, add a sound effect by adding a purple sound block named **play sound Cat Meow until done**. Change the audio file to your choice (the play again My Block is using **Baa**). Under this block, add a purple light block named **turn on** *smiley face* **for 2 seconds**. Click the face and change the graphic to your choice. In this example, an **X** was made. Add another light block underneath this one turning off all the lights. Finally, add a final purple light block named **write Hello** and change **Hello** to **Left**.

5. You will follow a similar process for the **define start game** block. Locate this pink block. Add a purple sound block named **play sound Cat Meow until done**. Change the audio file to one of your choice (in the example, it has changed to **Coin**). Add a purple light block named **turn on** *smiley face* **for 2 seconds**. Change this block to the number 3 in the graphical interface and change it from **2** seconds to **1** second. Right-click the purple **play sound** you just added and choose **duplicate**. Duplicate these two blocks two times so you don't have to keep dragging blocks. This process will create a visual countdown timer. Lastly, create another pink **My Block** and name it **RPS**. Add this block to the end of this code.

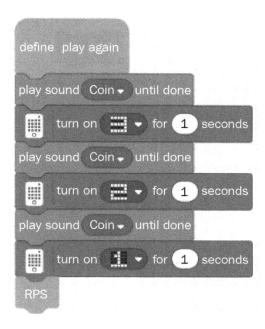

Figure 1.37 – start game My Block

The final step is to program the RPS block:

1. Add a purple **Light** block named **turn on** and change the smiley face to the four squares in the corners.

2. Add an orange **Control** block named **wait until**. Add a blue **Sensor** block of *Hub is shaken* to the diamond space.

3. Go to the orange **Variable** and make a variable named **Rps_random**. Add the **set Rps_random** block to the code. Use a green **Operator** block named **pick random** and select numbers 1-3.

4. Add an orange **Control** block named **if**. Drag in a green **Operator** block that compares with the = sign. On one side of the equal sign, add your **Rps_random** block and on the other side insert the number 1.

5. Within that **If** block, add a purple **Light** block named **turn on** and turn on all lights to symbolize paper.

6. Add one more purple **Light** block named **set Center Button light to** and choose a color.

7. Right-click this **if** block you just created and duplicate two more times. For these two copies, change the number to 2 and 1. Change the design from paper (all lights) to a smaller square for rock, and for the other design, a pair of scissors.

8. Add an orange **Control** block named **wait** and choose 5 seconds so you have time to see your game choice.

9. Add a purple **Light** block named **write Hello** and change it to **Right**.

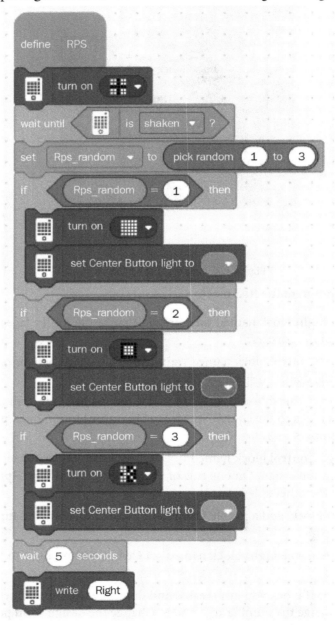

Figure 1.38 – RPS My Block

In the end, your code should look like this:

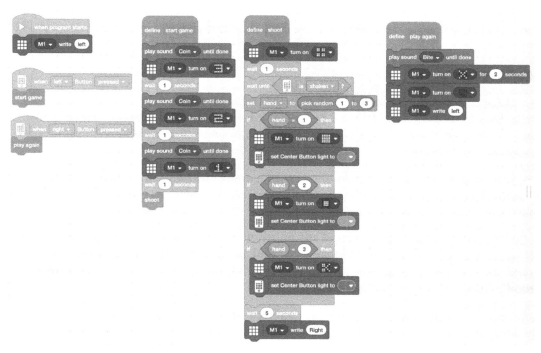

Figure 1.39 – Complete view of code

How it works is, the game will display **Left** on the screen to indicate to press the left button. This will trigger **Start Game**, which will do a countdown timer to show your choice. When it counts down, it will then move to the **RPS** block, where the code will wait for you to shake the Intelligent Hub before displaying a choice of rock, paper, or scissors. After it displays the choice for 5 seconds, it will then display **Right**, so when you press the right button, it will reset the screen and wait for you to press **Left** again to play the game.

Making your own game

Consider how you could take this simple and classic game and make it your own. Here are some suggestions that will allow you to use your own imagination and build up some skills as well:

- Design your own icons for rock, paper, and scissors.
- Convert this to a flip a coin (heads or tails) game.

- Convert this game from rock, paper, scissors to the rolling of a die. Can you rewrite the code so when you shake the Intelligent Hub it mimics the random roll of a die?

- Add an option where you can choose from the three games to play.

Summary

To wrap up this chapter, let's do a quick review of what we covered. We explored the overall look of the new LEGO SPIKE Prime kit. We examined the parts and elements that come with the kit along with the sensors. Additionally, we took some time to look at the programming software.

Now that we have a basic understanding of the kit and some featured elements, let's finally get to the building of robots. Let's take what we have learned and start building some fun, smart robots to challenge our thinking and creativity.

If you have not taken the time to build any of the other robot builds that are in the software, I would encourage you to take some time to build them and explore the ideas shared. None of these builds are required for the builds in this book, but they do provide a good foundation for the parts and building.

A few of these builds will reference certain builds to showcase how these ideas inspire new ideas.

In the next chapter, we will start with a robot arm and claw, a classic build that is always great to build to explore the parts, programming, and use of sensors.

2
Building an Industrial Robot Claw

Industrial robots have been around for a long time. In many cases, industrial robots are not the humanoid-looking robots we imagine when thinking of robots. Instead, many are robot claws that can do a wide variety of tasks, such as surgery, welding, assembly, painting, and more. In this chapter, you are going to build a robotic claw to pick up an object to better understand how these claws work and operate. In the following image, you can see what you will be building:

Figure 2.1 – A robotic claw

In this chapter, we will break down the build and program, as follows:

- Building the base
- Building the Intelligent Hub frame to move multidirectionally
- Building the robot arm
- Building the cargo
- Writing the code
- Making it your own

Technical requirements

For the building of the robot, all you will need is the **SPIKE Prime** kit. For programming, you will need the **LEGO SPIKE** application/software.

Access to the code can be found here: `https://github.com/PacktPublishing/Design-Innovative-Robots-with-LEGO-SPIKE-Prime/blob/main/Chapter%202%20-%20Robot%20Arm%20Code%20.llsp`.

You can find the code in action video for this chapter here: `https://bit.ly/3xkt1Mj`

Building the base

Before we build this **claw**, let's explore the strategy being used for this claw. There are many **LEGO** robotic claws to be found online. This claw will use the following tactics and strategies:

- The claw needs to be able to be controlled by a human.
- The claw needs to be able to move from side to side.
- The claw needs to be able to move up and down.
- The claw needs to be able to open and close to grab various objects.

We are going to start this build using a large yellow *11x19* base plate as the main building foundation for this robotic claw. Let's take the following steps:

1. Ensuring that we have a solid foundation is key, and the yellow *11x19* base plate shown in the following image is perfect for building on top of when designing a claw:

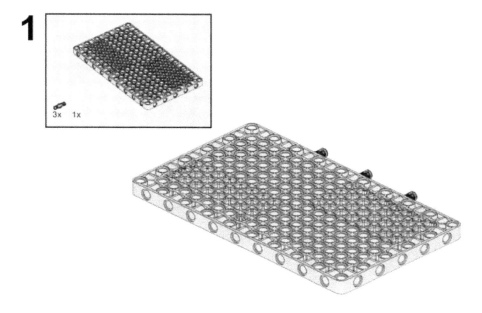

Figure 2.2 – The LEGO yellow base plate

2.   We are going to need to add the second yellow *11x19* base plate:

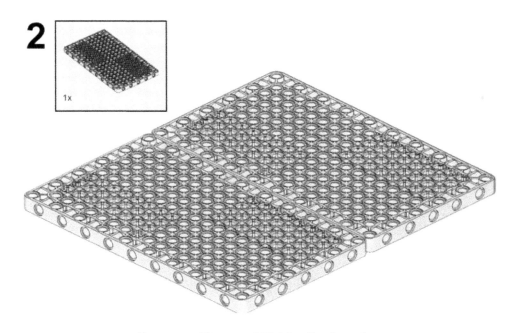

Figure 2.3 – The second LEGO yellow base plate

3. On each corner of the yellow base plates, you will add a gray perpendicular connector piece, as shown in the following figure:

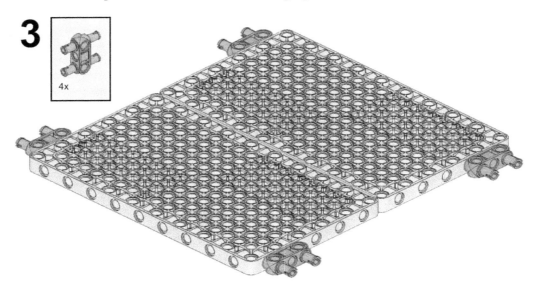

Figure 2.4 – Add gray connectors to edges

4. Next up, you will now add six *3L black axle* pins with friction ridges to the sides, as follows:

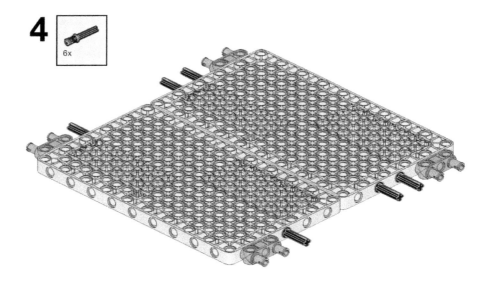

Figure 2.5 – Three axles added to each side

5. Now that we have plenty of connectors in place, it is time to build out the next part of the base, which will provide spaces for you to move an object from one location to another. Start by adding a *yellow 3L beam* and *purple 11L beam* to each side, as shown in *Figure 2.6*:

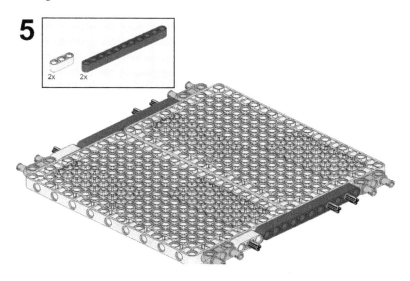

Figure 2.6 – Fill the space between the gray perpendicular connectors

6. From this step, attach an azure *11x15* open center frame on each side. Hold this piece in place using an azure *7L* beam and one gray perpendicular connector on each side.

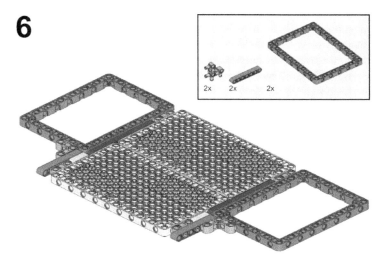

Figure 2.7 – Attach the open frames for object delivery

7. The final part to assemble on the base is the beginning space for your object. Begin by first adding two gray perpendicular connector pins to the front of the yellow base plate. Attach a black *3x11* panel plate to each connector. Next, attach the 8 black connector pins along this whole edge, as shown in *Figure 2.8*:

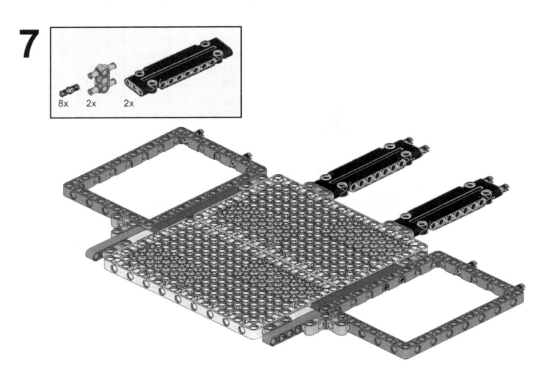

Figure 2.8 – Adding the black 3x11 panels

8. The next set of elements will help hold everything together. Attach an azure *13L* beam across the two black *3x11* panels using the black connector pins already in place. Insert three black connector pins to this beam, as shown in *Figure 2.9*.

Follow the same process for the two black *15L* beams across the azure open frames.

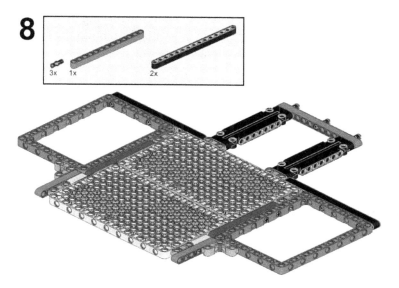

Figure 2.9 – Building the support for the front of the base

9. The final step to this section requires you to add two yellow *3L* beams to either end of the azure *11L* beam. Lock the black *5x7* open center frame to the azure beam using the black connector pin still available and using two more gray perpendicular connectors, as shown here:

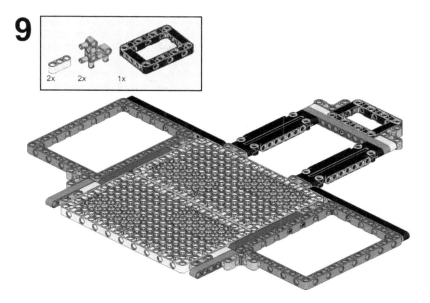

Figure 2.10 – The final look of the base

You now have completed the base for the robot arm and the playing field to move an object from one open frame to the other. Our next step is to build the frame for **Intelligent Hub** to serve as our arm controller.

Building the Intelligent Hub frame to move multidirectionally

One part of this kit that is different from previous **LEGO MINDSTORMS** kits is that the wires for all sensors and motors are set to a specific length. In previous kits, we could attach various cable sizes as we built our bigger structures.

Because all wires are a set length that we cannot adjust, we must consider this creative constraint in the build design. With that being said, we need to build a frame to place Intelligent Hub at a certain height to allow the motors to be able to reach it while in motion.

In this case, you will build a base that will provide adequate height along with an opportunity for Intelligent Hub to spin and pivot up and down. Let's look at the following steps:

10. To begin this aspect of the build, begin by securing two black biscuit elements to the yellow base plate using the blue connector pins. These are centered towards the back of the build and will be three pin holes from the edge. The following figure illustrates this:

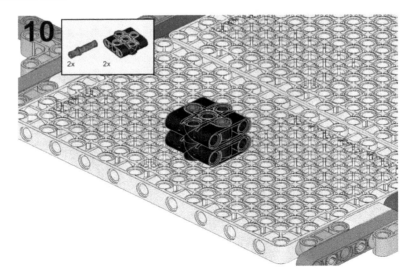

Figure 2.11 – Biscuit elements added to the yellow base plate

11. Next, attach the yellow *7L* axle to the middle of the two black biscuit elements. Slide the tire through the yellow axle. On top of the tire, insert two blue connector pins in the top and bottom pin holes and secure another purple biscuit, leaving space on top to connect more elements, as shown in *Figure 2.12*:

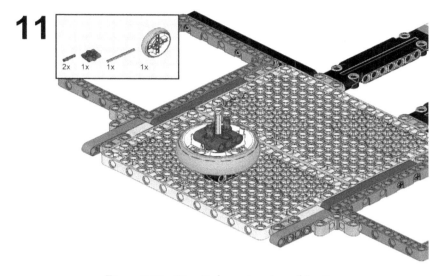

Figure 2.12 – Biscuit element on top of the tire

12. Once you have ensured everything is secure and connected, add one more tire to the top of the purple biscuit element, as shown in *Figure 2.13*:

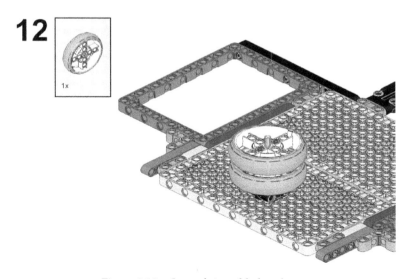

Figure 2.13 – Second tire added to the top

13. Just like the previous step, you will add another purple biscuit element to the top tire using two blue connectors pins but using the left-side and right-side pin holes, as shown in *Figure 2.14*:

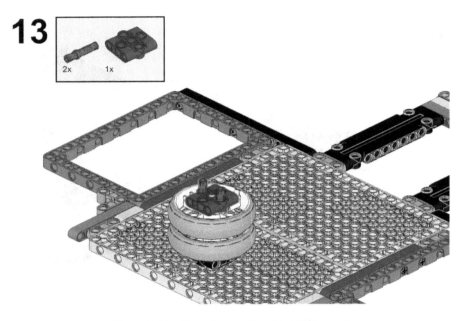

Figure 2.14 – Securing another biscuit element

This next sub-model will sit on top of the two tires we just installed. This will serve as the housing unit for Intelligent Hub, allowing it to move forward and backward while using the tire build and spin right and left:

14. To begin, locate a purple *7x11* open center frame and install four blue connector pins, as shown in *Figure 2.15*:

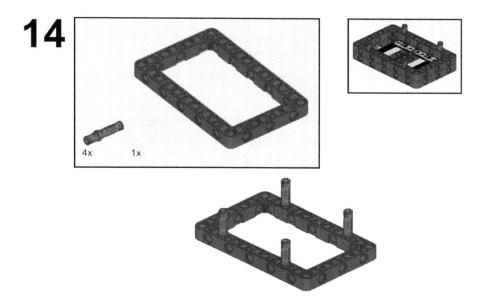

Figure 2.15 – Add four blue connector pins to the purple 7x11 open frame

15. Attach another purple *7x11* open frame using the blue connector pins, as shown in *Figure 2.16*:

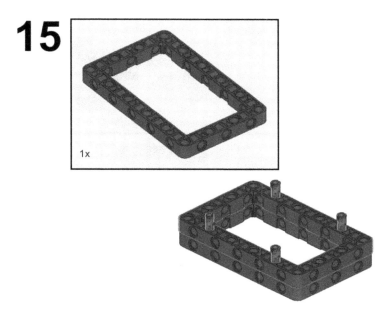

Figure 2.16 – Stack another purple 7x11 open frame

Keep in mind that, for this next part, I suggest you build the black open frame piece first. As you look at *Figure 2.17*, you can see how it fits within the purple *7x11* open frame, but it is not locked in at this point.

16. For this step, there is a three-image process to help you see where all the pieces go in *Figure 2.17*:

   A. Attach a yellow 3L beam to both sides of the purple biscuit. Connect the yellow 3L beams using two black connector pins on the outside pins.

   B. Insert this build into the black 5x7 open frame.

   C. Take this piece and add a gray pin with a bush stop to the middle pin hole on both *5L* sides of the black *5x7* open frame. This will allow these pieces to pivot back and forth.

Figure 2.17 – Piecing the elements together

And here is how it fits into the purple 7x11 open frame:

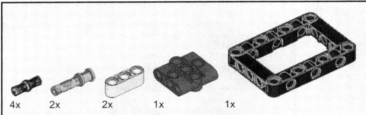

Figure 2.18 – Pivot piece for Intelligent Hub

17. This piece will fit into the purple open frame but, as you can see, when building, it does not stay together, but for a point of reference, these will sit atop the wheels, as shown in *Figure 2.19*:

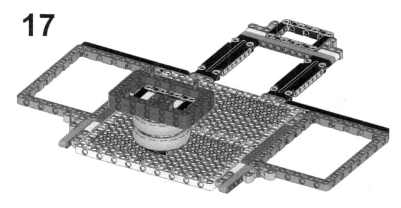

Figure 2.19 – Location of the piece on tires

18. Lock this piece into the purple open frame by using four more gray bush stop pins and connect two on each side, as shown in *Figure 2.20*:

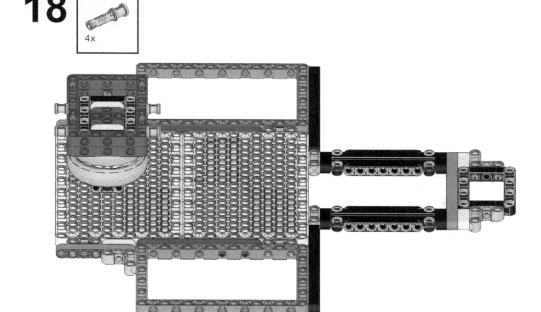

Figure 2.20 – Pins with bush stops hold it all together

19. Add a tan pin connector to each of the gray connectors on the bush stop axle insert, as shown in *Figure 2.21*:

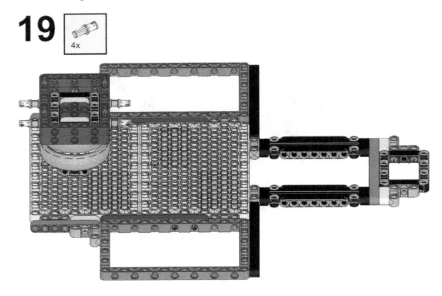

Figure 2.21 – Add tan connector pins to each bush stop

20. We will now add the **azure curved plates** to each side to act as a joystick mechanism to help steer Intelligent Hub back and forth, as illustrated in *Figure 2.22*:

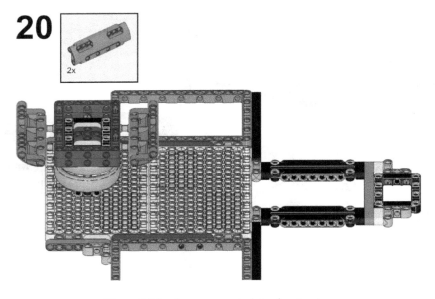

Figure 2.22 – Azure curved plates for steering

21. Finally, the last step for the controller of our robot arm is to add the most important piece, the Intelligent Hub, as shown in *Figure 2.23*. The Intelligent Hub connects to the four blue connector pins in place in the purple open frame. Click all into place and spin your Intelligent Hub around to ensure it properly spins on the yellow axle. Additionally, it should also be able to lean forward and backward.

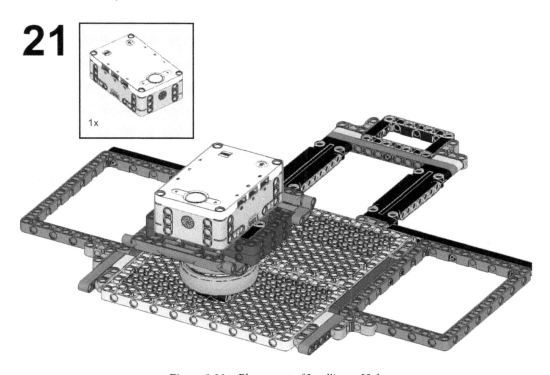

Figure 2.23 – Placement of Intelligent Hub

When you have tested these two movements, it is time to build the robot arm itself.

Building the robot arm

You are now ready to build the **robot arm** that will be controlled by Intelligent Hub. In order to achieve this, you will need to take the following steps:

22. Start with four black connector pins and add them to the front yellow base plate, as shown in *Figure 2.24*:

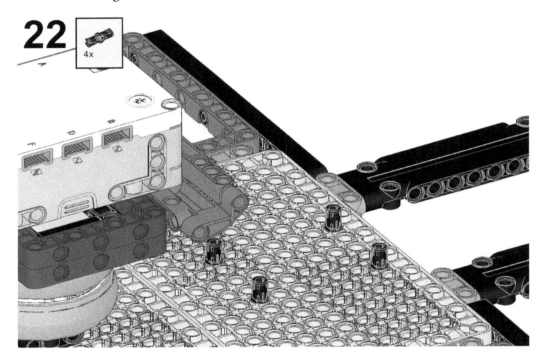

Figure 2.24 – Placement of the black connector pins

23. On top of these four black connector pins, add the large motor from the kit, as shown in *Figure 2.25*:

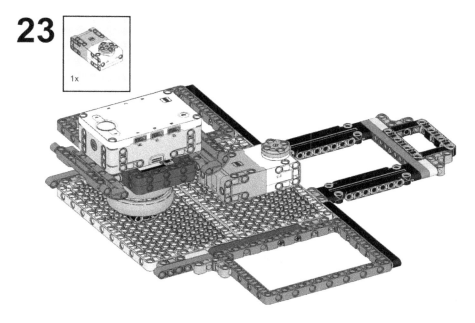

Figure 2.25 – Attach the large motor to the black connector pins

24. Once your motor is secure to the yellow base plate, add two blue connector pins and one black biscuit to the motor. The blue connector pins will be placed on the top and bottom pin holes of the motor, not the right and left sides.

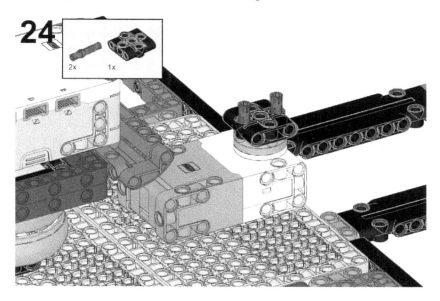

Figure 2.26 – Attach the blue connector pins and biscuit

Be sure your motor is aligned to position *0* (gray dots aligning):

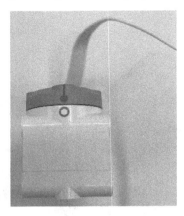

Figure 2.27 – Motor aligned to position 0

25. On top of the black biscuit element using the blue connector pins, you will add a medium motor facing to the left. Again, align the motor to position *0* (dots lined up as shown in *Figure 2.27*):

Figure 2.28 – Attach the medium motor

26. This next part of the build is to create a simple stopper, so the robot arm does not go straight back and hit Intelligent Hub.

Begin by adding four black connector pins to the pin holes on the medium motor in the white pin holes. Attach two azure *7L* beams facing up, as shown in *Figure 2.29*:

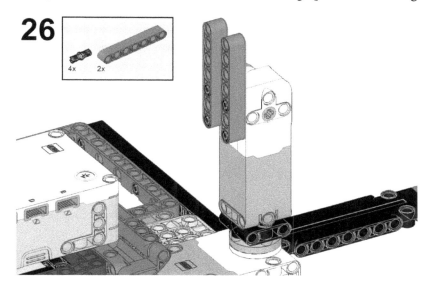

Figure 2.29 – Attach the azure 7L beams

27. Add a yellow *2x4 L* beam to the top of each azure *7L* beam using two black connector pins for each one, as shown in *Figure 2.30*:

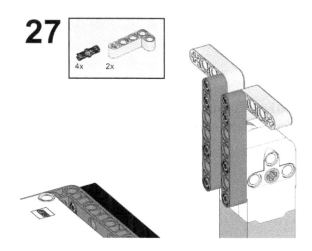

Figure 2.30 – Attach the yellow 2x4 L beams

This has been designed to be simple so that as you tweak this build on your own, you can easily remove these pieces if you wish.

It is now time to add the second motor for the robot arm by taking the following steps:

28. Begin by adding two tan connector axle pins to the middle pin hole of the medium motor on either side. On the side of the motor with the azure motor element, add one black connector pin to the top pin hole, as shown in *Figure 2.31*:

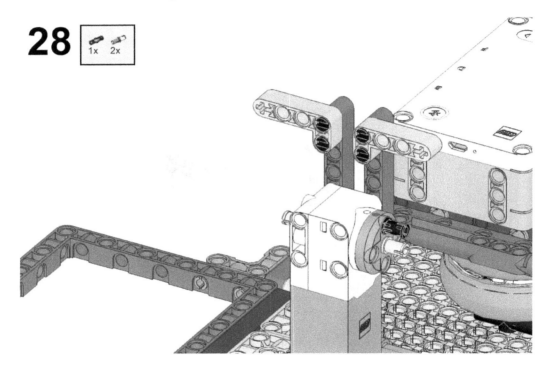

Figure 2.31 – Prep for the claw

29. We are now going to build a **sub-model** of the robot claw that will fit onto the connector pins we just added to the medium motor.

Start with one black *15L* beam, as shown in *Figure 2.32*:

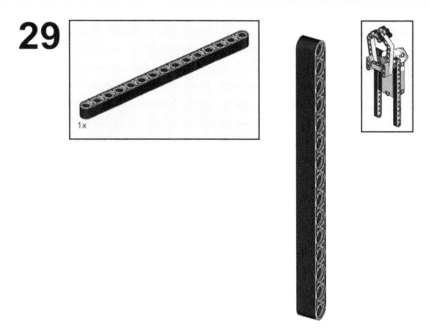

Figure 2.32 – Beginning of the claw build

30. Using two blue connector pins, attach a black *9L* beam to the *15L* beam, as shown in *Figure 2.33*:

Figure 2.33 – Attach the 9L beam

31. Using the two blue connector pins, add the second medium motor, as shown in *Figure 2.34*:

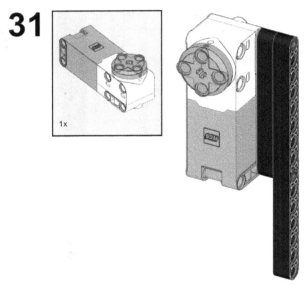

Figure 2.34 – Attach the second medium motor

32. On the other side of the medium motor, which does not have anything added yet, use two black connector pins and attach another black *15L* beam. Let's see what this looks like in an illustration:

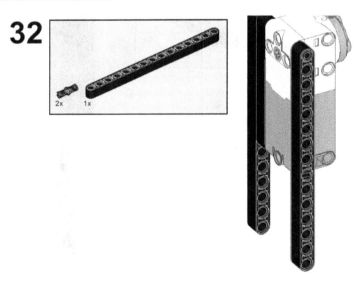

Figure 2.35 – Attach another black 15L beam

33. On the backside of the motor, insert a yellow *3L* axle and two black connector pins to the right and left sides, as shown in *Figure 2.36*:

Figure 2.36 – Attach a yellow axle and pins

34. Slide a purple *5L* beam across the pin connectors. Add two more black connector pins to each end of the beam, as shown in *Figure 2.37*:

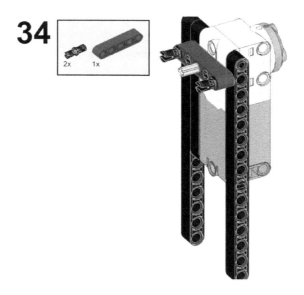

Figure 2.37 – Attach a purple 5L beam

35. This next little build is what will allow your claw to open and close. Add a tan axle connector pin to a white axle and insert a pin connector element into the right axle hole:

 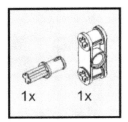 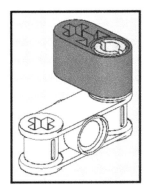

Figure 2.38 – Claw movement mechanism

36. Add an Azure *1x2* axle hole element to the pin side of the tan axle connector, as shown in *Figure 2.39*:

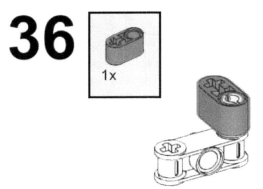

Figure 2.39 – Add the 1x2 axle hole

37. Attach this build element to the backside of the motor, as shown in *Figure 2.40*:

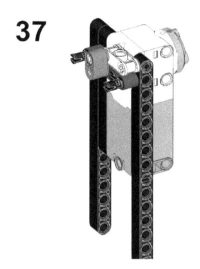

Figure 2.40 – Insert into the medium motor

38. Using the two black connector pins that are still open on the purple beam, add one yellow bent **lift arm** to each pin, as shown in *Figure 2.41*:

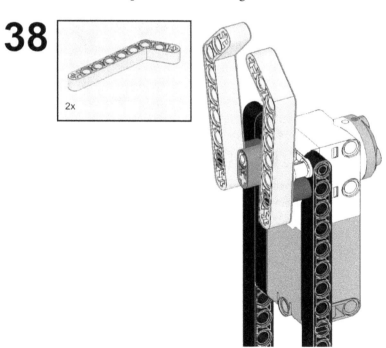

Figure 2.41 – Add the yellow lift arms to the motor

39. If looking at the build straight on, you will add one black axle connector pin on the right side in the third pin hole from the bottom. This will connect the yellow lift arm to the white connector piece behind it. You will add one tan axle pin connector to the left yellow lift arm in the bottom hole. Let's see what this looks like in *Figure 2.42*:

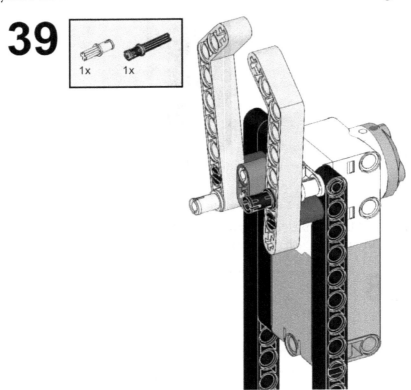

Figure 2.42 – Add tan and black connector pins

40. Secure the yellow lift arms together for movement to occur by using one H-shaped lift arm to hold them in place:

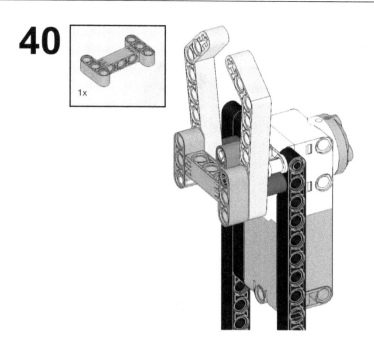

Figure 2.43 – Add a H-shaped lift arm

41. The final step is to add some grippers so that the arms can hold an object without slipping. Add a yellow *3L* axle through the end of each yellow bent lift arm, as shown in *Figure 2.44*:

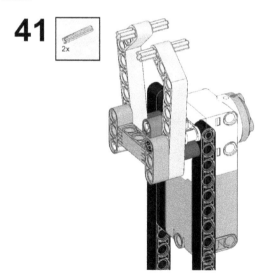

Figure 2.44 – Add a yellow 3L axle beam

42. Add a rubber gripper to each side, as shown in *Figure 2.45*. You will use four of them in total:

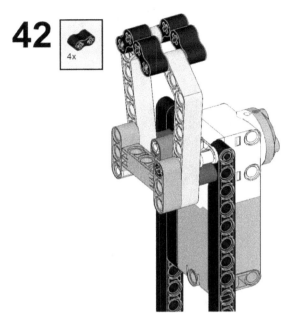

Figure 2.45 – Add grippers

43. The final step is to now add this claw to the overall **build design**. Use the pins on the first medium motor to connect everything into place. Let's look at an illustration of this:

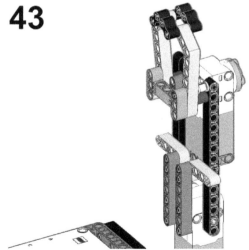

Figure 2.46 – Add the claw to the robot

44. And you now have one sweet-looking robot arm build, ready to be programmed and brought to life! Let's have a look at it in *Figure 2.47*:

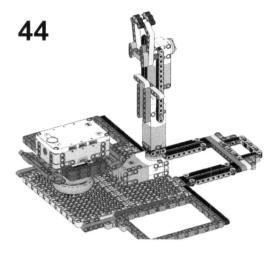

Figure 2.47 – Final view of the robot

Now that your claw is complete, we have one more item to build, which is an element to be picked up, so let's build a small **cargo element**.

Building the cargo

Now that the arm mechanism has been built, you need an element to pick up. You will now build a cargo build to practice using your claw:

1. Start with two black *2x8* plates and place them side by side, as illustrated in *Figure 2.48*:

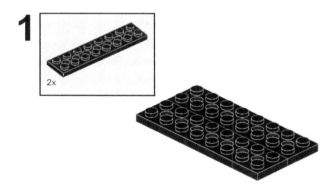

Figure 2.48 – Two black 2x8 bricks

2. Connect these elements together using two purple *2x5* blocks and place them at each end, as shown in *Figure 2.49*:

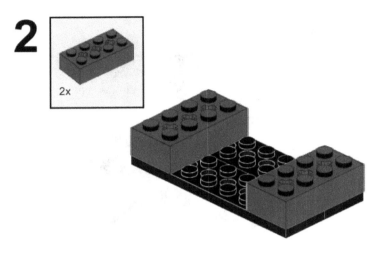

Figure 2.49 – Join together using 2x4 blocks

3. Insert a yellow *3L* axle into the middle of each purple *2x4* block, as shown in *Figure 2.50*:

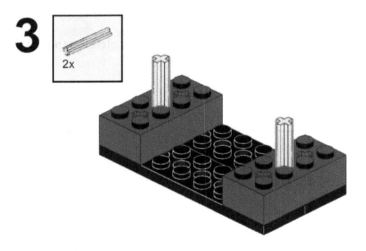

Figure 2.50 – Insert a yellow 3L axle

4. Slide a gray bushing onto each yellow axle, as shown in *Figure 2.51*:

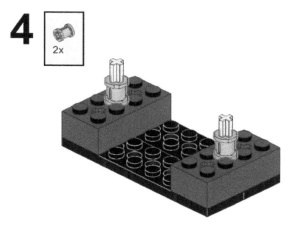

Figure 2.51 – Insert a gray bushing into each axle

5. The next step is to add a yellow 90-degree connector to the top of each yellow axle, as shown in *Figure 2.52*:

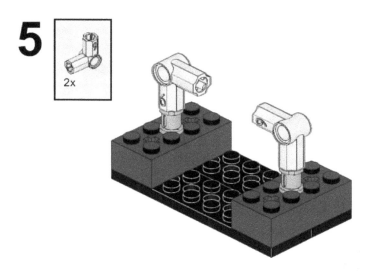

Figure 2.52 – Add a yellow 90-degree connector

6. The final step is to use a *5L* axle and, in the middle, slide it onto a tire. Secure the tire using a gray bushing on each side. Finally, work the *5L* axle into the two yellow 90-degree connectors, as shown in *Figure 2.53*:

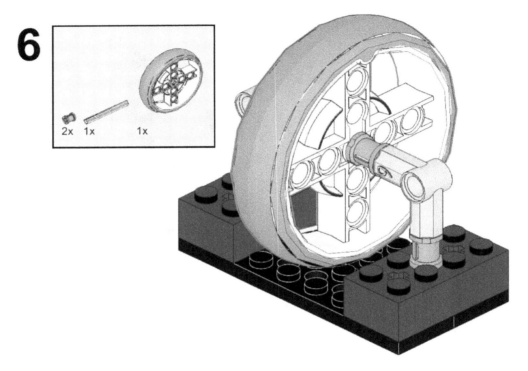

Figure 2.53 – Add the tire to the middle of a 5L axle

This will be the cargo item we will use to grab and move to different locations. Next, it will be time to move on to the coding to bring the robot claw into use.

Writing the code

For this build, we are going to focus on **writing code** that allows us to use the Intelligent Hub as a remote control for the arm. This will be a program that allows you to control the robot using the gyro sensor feature of the Intelligent Hub to move the claw left, right, up, and down, and to open and close the claw.

Overall, this program is simple to develop and is created to be easily adaptable for you to remix to meet your needs. Use this sample code to make sure everything works, and then begin to tweak the code to make it work to your desired needs. Make sure you have the proper ports plugged in, and then move on to the code.

Identifying the ports

If you have not plugged in your motors yet, then let's get them plugged into the proper ports. You will plug the large motor that moves the claw to the right and left to motor **port E**. The medium motor that controls the opening and closing of the claw will plug into **port A**. The medium motor that controls the arm of the claw to go up and down will use **port C**. Once you have the motors plugged in, then double-check that the motors are plugged in correctly.

You can check that your motors are properly plugged in by checking the **Hub Connection** in the software, as seen in the following screenshot:

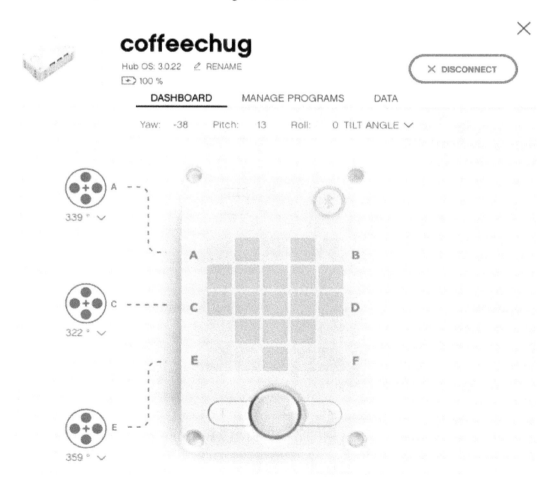

Figure 2.54 – Hub Connection view

It is time to write the program so that you can control the claw.

The Intelligent Hub remote-control robot program

To get started with this program, let's go ahead with the following steps to make the robot come to life:

1. Open up the **SPIKE Prime** software.

    Click on **New Project** at the center of the main screen:

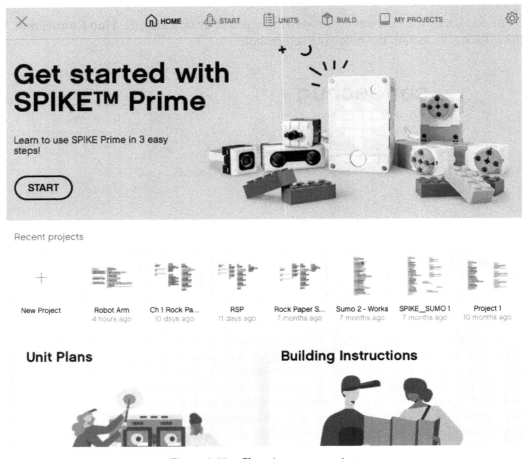

Figure 2.55 – Choosing a new project

2. On the **New Project** window, select the **Word Blocks** option, give your program name, and click on **Create**.

3. When the coding canvas opens, there is already a default **when program starts** block provided, as shown in *Figure 2.56*. Let's begin with this **code stack**. We will add some code to align all our motors to a common starting point at the beginning of the code. Add a **cyan sensor** block called **set yaw angle to** under the **when program starts** block:

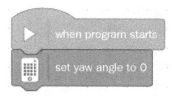

Figure 2.56 – Creating the initial code stack

4. **Yaw** is the orientation of Intelligent Hub turning right or left, as shown in *Figure 2.57*. In other words, it is the rotation around the z axis:

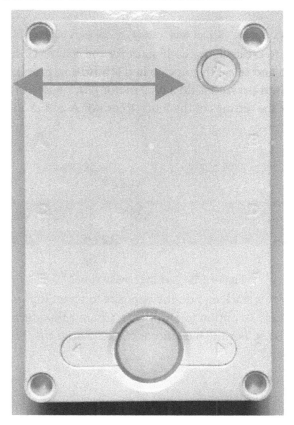

Figure 2.57 – Yaw is turning Intelligent Hub left and/or right on the z axis

5. The last two blocks we will add to this stack are blue **motor** blocks. Add two of the motor **go shortest path to position** blocks, as shown in *Figure 2.58*. Set one block for motor **E** and the other for motor **C**:

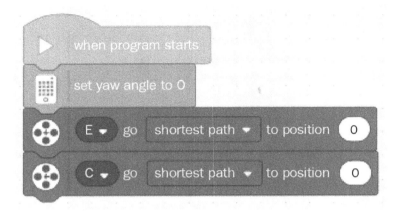

Figure 2.58 – Set motors E and C to the start position

We are now going to create a new code stack for the claw to close. Go to the **Events** code block section and select the block **when left Button pressed**. Change this block to **right Button pressed**. Add a blue motor block named **run for 1 clockwise rotations**. Change the settings of the block to motor **A** and change the settings to **7 degrees**:

Figure 2.59 – Set the claw to close

You will now copy this stack and do the opposite to open. Right-click the **Event** block and you will see an option to **Duplicate**. Change this stack to **left Button** on the orange event block. For the blue motor block, change the direction of the turn to *counterclockwise*:

Figure 2.60 – Set the claw to open

6. The final code stack will be created to allow the arm to move in the direction of Intelligent Hub using angle and yaw. Begin by adding an **Event** block named **when program starts**.

7. Add a blue motor block to adjust the motor **C** speed to 5%. You can adjust the speed to your liking once you get the arm moving.

8. The next block we will add is a **Control** block named `forever`. In this block, add another blue motor block, moving motor **E** to the shortest path to position the block. Where the number 0 sits, you will add a cyan sensor block named `Pitch Angle`. Add that to 0 and then change the pitch to **Yaw**. *Figure 2.61* will showcase what you should have at this point:

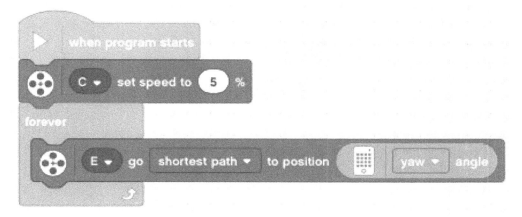

Figure 2.61 – Controlling arm code

9. Next, we will create an `If Then` statement and basically duplicate it two more times to complete the code. This will program the arm to follow the motion of Intelligent Hub. Go to the **Control** blocks and add an **If Then** block into the `Forever` block of our code. Add a green **Operator** block of `less than`. On the left side, add a cyan **Sensor** block of **Pitch Angle**, and to the right of the equation, add `-3`. Under that block, add a blue motor block, **Start Motor**, and change to motor **C** and in a counterclockwise direction:

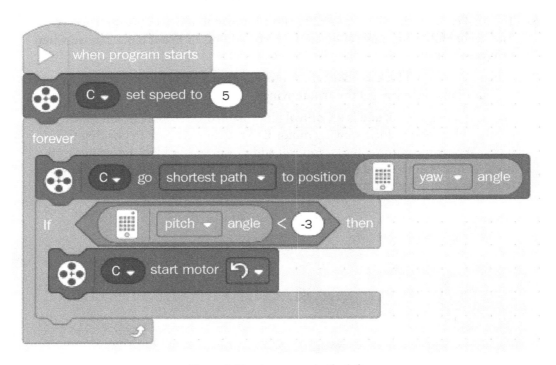

Figure 2.62 – Arm move to the left

10. We will now copy this format, but use different **Operator** blocks for greater than and another **Operator** block when the pitch is between -3 and 3. See *Figure 2.63* for the rest of the code to complete the sequence:

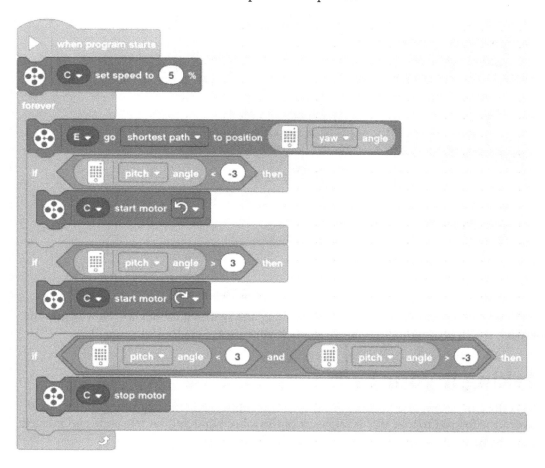

Figure 2.63 – Complete code for arm control

When it is all complete, your code should look like this:

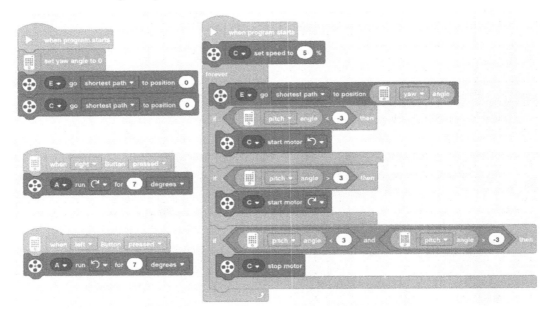

Figure 2.64 – Complete code for this project

And that is it! You did it. You built a nice-looking and working robot claw. Now comes the fun part, which is to remix the claw to **make it your own**. Let's look at some ideas.

Making it your own

This is the part I love, and I hope you do too. This robot is just the beginning of the fun. It is now time for you to take this framework of the robot build and sample code and make it unique to your talents and imagination.

Here are a few ideas to consider applying to this robot:

- Add sensors to trigger autonomous robot decisions. For example, could you add a color sensor so that, as the claw moves, it is waiting to detect a color to stop and pick up the object and move to another location?
- Could you program the arm to move without human interaction?
- Instead of using the buttons on Intelligent Hub to open and close the claw, could you use the Force Sensor instead?
- How could you modify the claw to grab different objects and sizes?

- Add lights from the **Light-Emitting Diode** (**LED**) matrix on Intelligent Hub to provide insights or cool new looks as the robot operates.

- Add sound effects to make it sound like an industrial robot.

Summary

Industry continues to rely on robotics to achieve production goals. By the end of this chapter, we had explored the concept of industrial robots by building a robotic claw that can pick up an object and transport it to another location. Additionally, we explored how to have several motors working together to create an actual working claw. This is a classic build that must be done with any new robotic kit!

Let's now head to the next chapter and explore a whole different world of robotics by building a guitar.

3
Building a LEGO Guitar

Entertainment is an important aspect of our lives and well-being. One of the great hobbies, interests, and careers is music. As technology continues to advance in our lives, we see more technology being used to create musical, music experiences, concerts, and so on.

This chapter will be combining robotics and music by using the **LEGO SPIKE Prime** kit to make a guitar that is playable and codable to our unique needs.

The gaming industry was impacted by the awesomeness of *Guitar Hero* and the likes of many other music games; let's create our own instrument that will allow you to jam to your favorite song.

Here is what your guitar will look like by the end of this chapter:

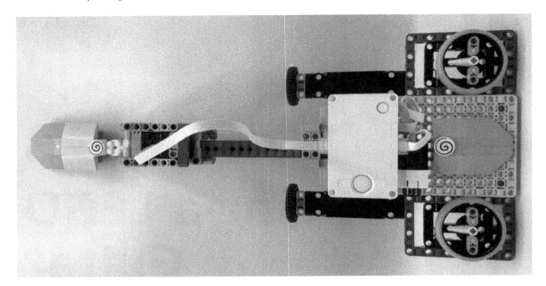

Figure 3.1 – The final guitar build

In this chapter, we will break down the guitar build and coding into the following sections:

- Building the guitar
- Coding the guitar
- Playing the guitar
- Making it your own

Technical requirements

For the building of the robot, all you will need is the **LEGO SPIKE Prime** kit. For programming, you will need the LEGO SPIKE Prime app/software.

Access to the code can be found here: `https://github.com/PacktPublishing/Design-Innovative-Robots-with-LEGO-SPIKE-Prime/blob/main/Chapter%203%20-%20Guitar%20Code.llsp`.

You can find the code in action video for this chapter here: `https://bit.ly/2ZhzxXl`

Let's start building it!

Building the guitar

You will build the guitar in sections. Each section will provide a basic framework for the guitar, but please keep in mind that with all the builds, you have the space and opportunities to build it the way you want. In terms of the guitar, the key pieces to customizing the guitar will be the body of the guitar, the top of the neck of the guitar, and the LED lights. Additionally, when you get to the coding, you can really fine-tune it to have the guitar sound just the way you want it to sound.

Let's get started with the building.

Assembling the guitar body

This is where you get to spice up the guitar to your liking. For the sake of this example, you will build a basic outline of a guitar, but please know that from here you can design it to your own liking:

1. Begin by adding four black connector pins to each of the corners of a yellow *11x19* base plate:

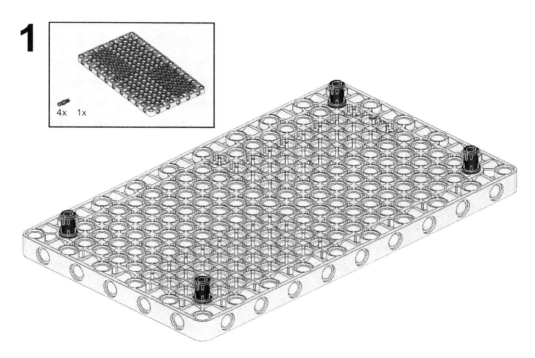

Figure 3.2 – Black connector pins on the base plate

2. Attach the second yellow *11x19* base plate on top and then add four black connector pins to each side of the bottom yellow *11x19* base plate:

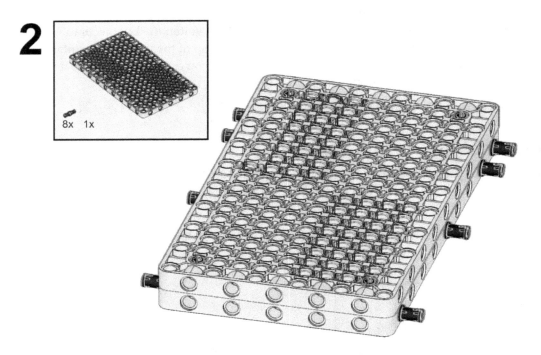

Figure 3.3 – Black connector pins and second base plate added

3. To either side of the bottom yellow base plate, connect a purple *7x11* open frame:

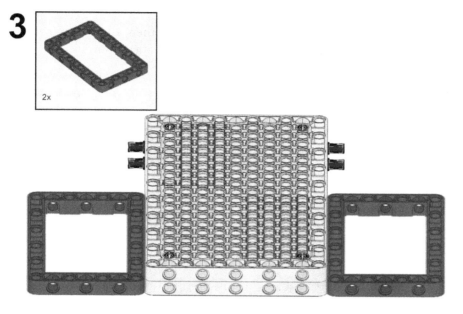

Figure 3.4 – Purple open frames added to each side

4. Inside of each purple *7x11* open frame, add a black *5x7* open frame. To secure these to the purple *7x11* open frame, use two gray connector pins with bush stops. Add one to the opposite inside pin holes on the inner side of the black *5x7* open frames:

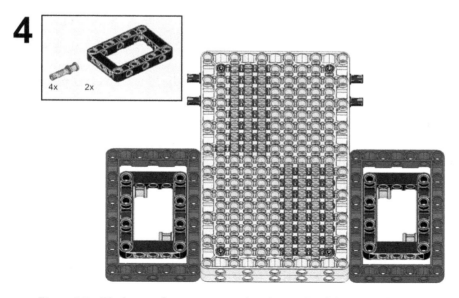

Figure 3.5 – Black open frames connected to the inside of the purple frames

5. Now that the open frames are secured to the base plate, it is time to add two blue connector pins to each of the black *5x7* open frames. Then, add a black connector pin to the top inner corner of each of the purple *7x11* open frames:

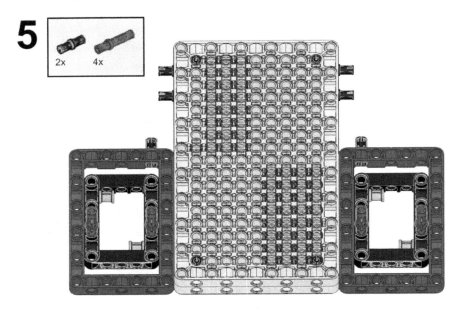

Figure 3.6 – Blue connector pins added to the black frames and a black connector pin added to the purple frames

It is now time to add some detail to the guitar body to make it look a bit nicer. Again, this is a wonderful place to add your flare, but here is one example of how to design the guitar.

6. First, begin by adding a wheel to each side, using the blue connector pieces available on the black *5x7* open frame.

    Second, add a black flat *3x11* panel to each side of the guitar, using the two black connector pins on each side of the yellow base plate and the one black connector pin on the top side of the purple open frame.

    Third, add a gray perpendicular four-pin piece to the top of each of the black flat *3x11* panels. At the end of each of these elements, attach a black 36-tooth double bevel gear:

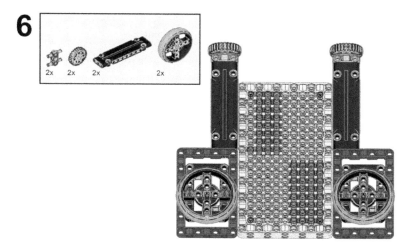

Figure 3.7 – Adding design elements to the guitar body

7. The final touch to the guitar body is to add a few last elements to the wheels. Start by adding two azure *1x2* beams with axle holes to each of the wheels, using the blue connector pin that you used for the wheels and supporting with a tan frictionless axle pin.

    In the middle pin hole between the two azure *1x2* beams with axle holes on each wheel, add a red axle pin and attach a white *1x3* tooth:

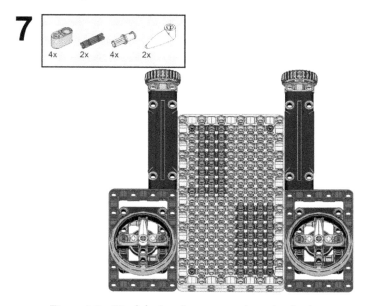

Figure 3.8 – Final design elements on the guitar body

8. Your next step is to add the Intelligent Hub. Using four black connector pins, attach the Intelligent Hub to the top of the yellow base plate:

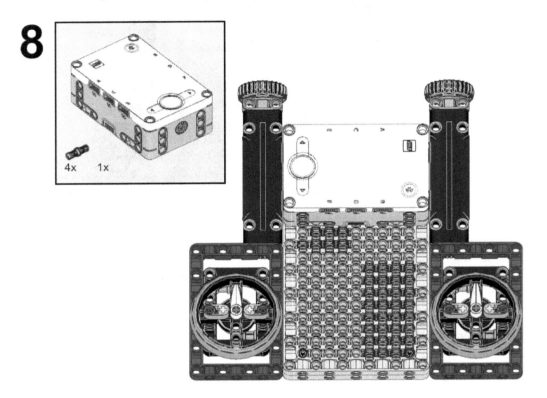

Figure 3.9 – Attaching the Intelligent Hub

9. Now that the Intelligent Hub is in place, it is time to add the force sensor, which will be used to trigger the notes of the guitar. Use four black connector pins to secure the force sensor. Use two black connector pins to secure the force sensor to the yellow base plate and two black connector pins to secure the sensor to the Intelligent Hub:

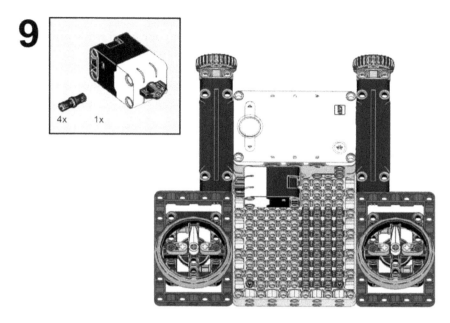

Figure 3.10 – Attaching the force sensor

10. Add a yellow *2x4* yellow brick right next to the force sensor, using two tan frictionless axle pins:

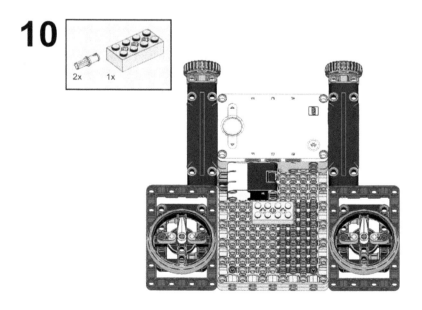

Figure 3.11 – Adding a yellow 2x4 brick

11. On top of the yellow *2x4* brick, secure two azure *8x3x2* wedges (one left and one right) to create a comfortable place for the palm of your hand while also providing some design. Secure those using a *2x2* round spiral pattern tile:

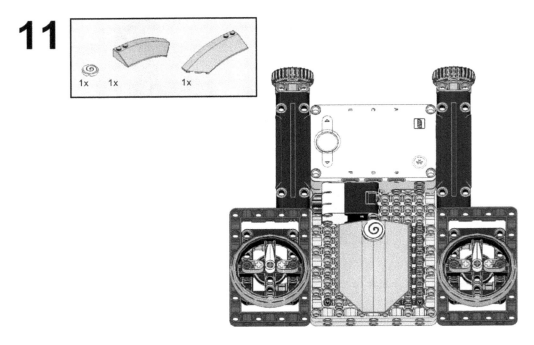

Figure 3.12 – Adding a spot for the palm of your hand

There is just one final build to complete the guitar body before you begin to build the fretboard. This piece will help secure the fretboard to the guitar body.

12. Secure two azure *3x5* L-shaped beams to the Intelligent Hub and yellow base plate, using two blue connector pins for each piece:

12

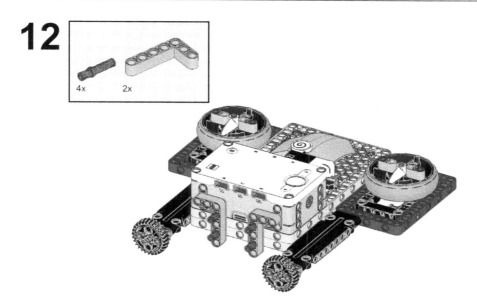

Figure 3.13 – Attaching 3x5 L-shaped beams

13. Finally, secure one gray *3x5* H-shaped beam to the top blue connector pins. Using two black connector pins, add another gray *3x5* H-shaped beam to the first one:

13

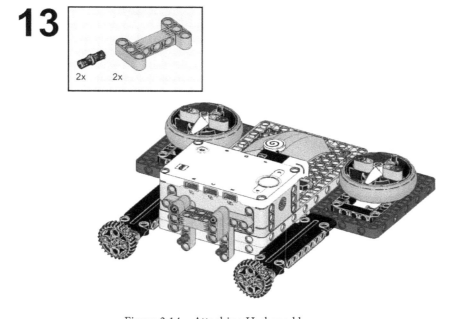

Figure 3.14 – Attaching H-shaped beams

Now that you have completed the guitar body, it is time to move to the fretboard of the guitar.

Assembling the neck

You will now build the neck of the guitar. For this, you are going to use the *2x4* colored bricks to create sections for the color sensor to detect. The color sections remind me of the *Guitar Hero* guitars, but instead of pressing buttons, we will use the color sensor to detect the colors.

You will build five different-colored sections for the fretboard. This maximizes the length of the color sensor wire and gives the player five different notes to play. Let's build one together:

14. Begin by using a purple *2x16* plate. Attach this plate to the top of a black *2x8* plate. Attach another purple *2x16* plate on the underside, aligned with the black *2x8* plate. Next, add an azure *2x2* round plate to the top of the build, along with another black *2x8* plate:

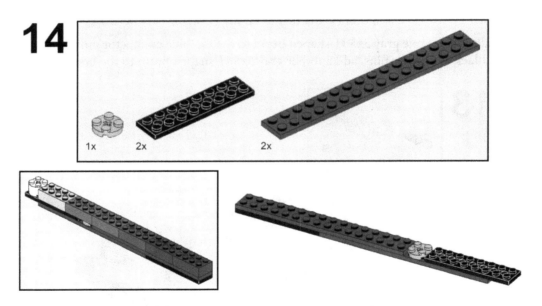

Figure 3.15 – The beginning of the fretboard

15. Attach the *2x4* colored bricks to the top of the fretboard. The key here is to be sure to use two of the purple *2x4* colored bricks to begin the fretboard. Additionally, at the end of the fretboard, attach a white *2x4* round brick:

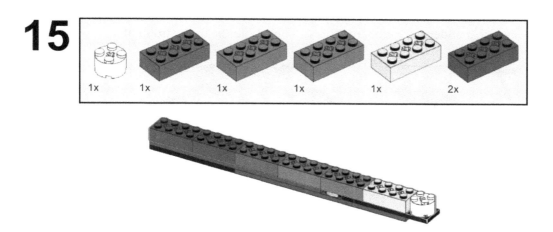

Figure 3.16 – Adding the colored bricks

16. Go ahead and attach the fretboard to the underside of the gray *3x5* H-shaped beams:

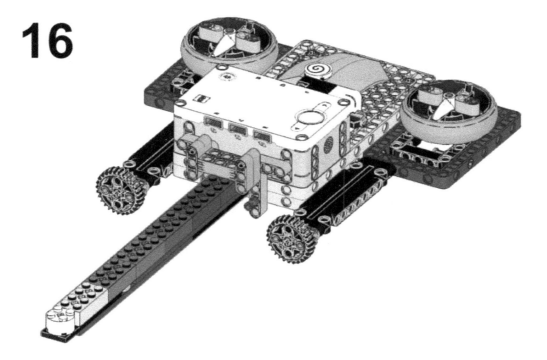

Figure 3.17 – Attaching the fretboard to the body of the guitar

17. Under the fretboard, attach one purple *5L* beam, using the blue connector pins still available under the fretboard. This will help hold the fretboard in place:

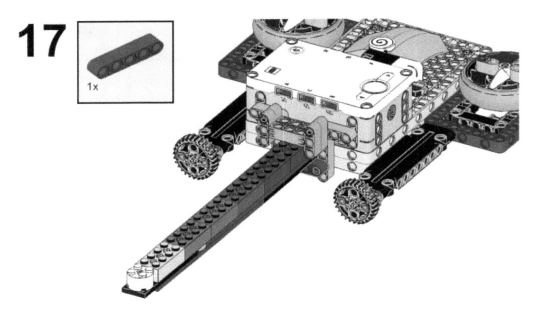

Figure 3.18 – Attaching purple beams underneath the fretboard

Now that the fretboard and body are complete, it is time to build the color sensor slider to be able to play the guitar.

Building the color sensor slide bar

The next piece we need to build is the slide bar, which will allow the color sensor to slide up and down the fretboard to read the colors of the *2x4* colored bricks you just put on the fretboard:

18. Let's begin by adding a purple *11L* beam to an azure *11x3* curved panel, using two blue connector pins:

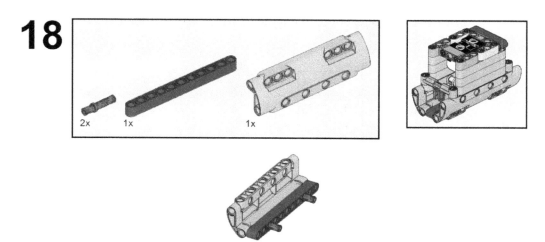

Figure 3.19 – Attaching the beam to the curved panel

19. Using the blue connector pins you just attached, secure another azure *11x3* curved panel on the other side of the purple *11L* beam:

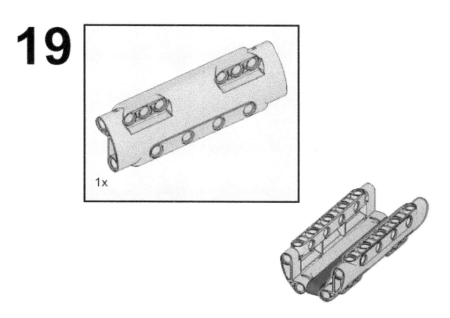

Figure 3.20 – Attaching another curved panel

20. On the inside of the right azure *11x3* curved panel, insert four tan frictionless pin connectors:

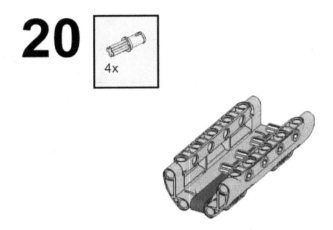

Figure 3.21 – Inserting four tan connector pins

21. To each of the tan connector pins, attach a wire clip. The colors are irrelevant so choose the colors you wish to use. These are added to secure the slider along the fretboard:

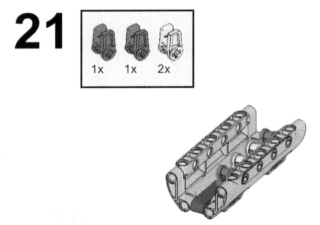

Figure 3.22 – Attaching four wire clips

22. On top of this build, let's secure the pieces together. Using four axle and pin connectors with an angle label #1 (a number 1 will be on the piece) and two yellow *3L* axles, build two parts to hold the parts together:

22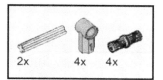

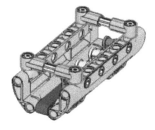

Figure 3.23 – Securing the build

23. Add two azure *7L* beams between the support pieces, using two blue connector pins for each beam:

23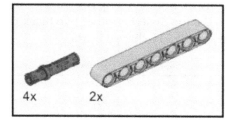

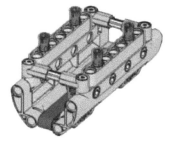

Figure 3.24 – Attach two azure 7L beams

24. Attach another layer of azure *7L* beams and add two black connector pins to each of the beams:

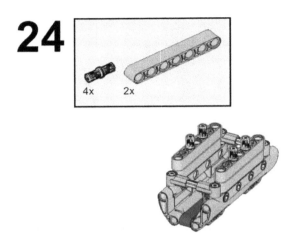

Figure 3.25 – Attaching another two azure 7L beams

25. Add a third layer of azure *7L* beams. Using two black connector pins, attach a gray *3x5* H-shaped beam to the top. Insert another two black connector pins to the inside pin holes of the gray *3x5* H-shaped beam:

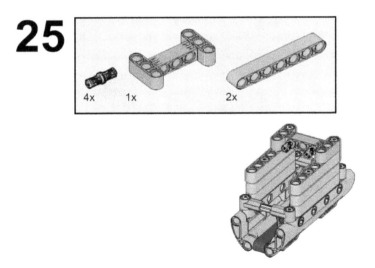

Figure 3.26 – Attaching a third layer of beams

26. Using the two black connector pins you just added to the gray *3x5* H-shaped beam in the previous step, attach the color sensor to these pins. On the other side of the sensor, insert a gray perpendicular four-pin piece, followed by a purple *5L* beam to the other side of this piece. Finally, secure two yellow *3L* beams on either side of the sensor, using black connector pins:

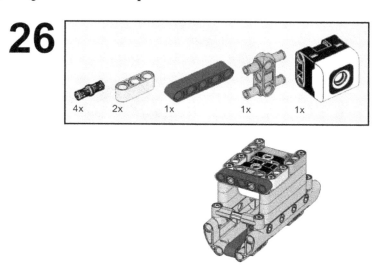

Figure 3.27 – Adding the color sensor

27. It is now time to slide this slider piece onto the fretboard, as shown in *Figure 3.28*:

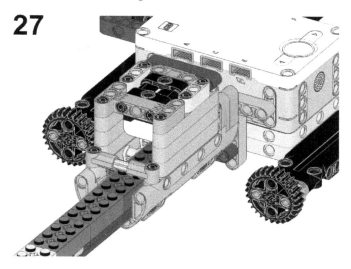

Figure 3.28 – Adding the slider to the fretboard

28. Your final steps are to complete the top of the fretboard with some design elements. Begin by adding a black *2x8* plate to the end of the fretboard. Attach a red *2x4* brick to the underside of this black plate, leaving two pegs open at the end. Add four yellow *3x2* curved slope pieces to the black *2x8* plate and red brick:

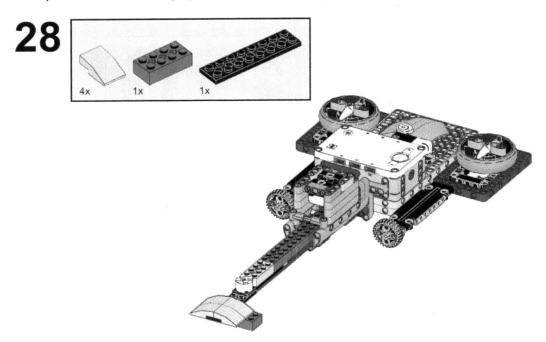

Figure 3.29 – Adding design to the end of the fretboard

29. On the underside of the fretboard, secure everything together using another black *2x8* plate with two black *6x2* curved slopes attached and an azure *8x6x2* windscreen element. The black *6x2* curved slopes connect to the underside of the red *2x4* brick and the azure *8x6x2* windscreen connects to the underside of the black *2x8* plate that is at the end of the fretboard:

Figure 3.30 – Adding end elements to the fretboard

30. The last step is to build a stopper for the slider so that it does not slide off and break the end of your guitar. In the last space on the fretboard next to the white *2x2* round brick, attach an azure *2x2* round plate with a white *2x2* round brick on top. Finally, add the *2x2* round tile with the spiral to the top of the white *2x2* round brick:

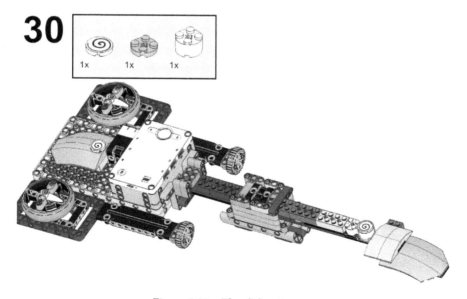

Figure 3.31 – The slider stopper

Now that our guitar is built, let's write the code for it!

Writing the code

The coding for the guitar is very simple in premise but allows for completely individual interpretation of how you want to play the guitar and how you want the guitar to sound.

The program we are writing as an example will showcase some possibilities, but in the end, be brave and tinker around to get the guitar to sound like you want it. The beauty of music is that it allows you to express yourself the way you want to express yourself. This is your moment! Combining coding and music is an exciting combination of awesome.

The ports

There is not a lot to plug in for this build. You will connect the force sensor into port D and plug the color sensor into port C:

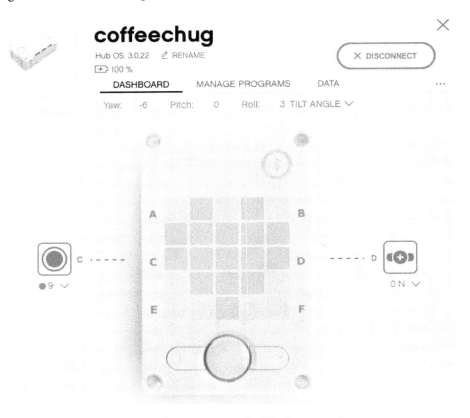

Figure 3.32 – The port view in the Mindstorms software

The basic layout of the program will follow this structure for each of the colored items on your guitar fretboard.

You will start by deleting the default yellow event block called **when program starts**. Right-click on this block and delete:

Figure 3.33 – Event block

Add a yellow event block named **when**:

Figure 3.34 – The when event block

In the coding canvas area, add a green logic **and** block:

Figure 3.35 – The and logic block

Inside each of the empty spaces of this logic block, you will add two conditions using the azure **Sensor** blocks. First, you will add the color sensor block and will set the color to the first color on your fretboard. In the model build, this color is purple:

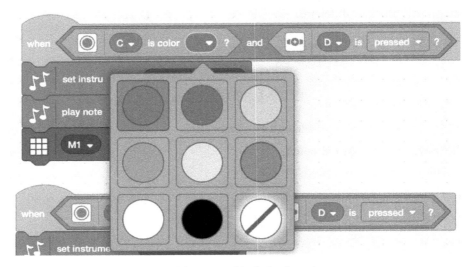

Figure 3.36 – The sensor block for the color sensor

The second sensor condition will be the force sensor:

Figure 3.37 – The sensor block for the force sensor

Next, you need to add some extension blocks to be able to code music. Click on the extension block icon at the bottom left of the programming menu. From there, install the **Music** blocks:

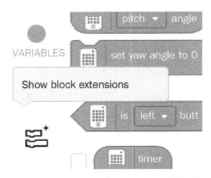

Figure 3.38 – The extension blocks located in the white outlined blocks at the bottom left of the screen

The music block extension provides you with a nice selection of choices to play music. You can experiment to have the guitar play the notes and music that you prefer. You can have a lot of fun tinkering around, tuning your guitar to the way you want it to play and sound:

Figure 3.39 – The music extension provides many options for playing music

In the next figure, you can see how the program comes together to play the notes for each color reading. I really liked the sound of the electric guitar. I found some basic sheet music for a song I knew I could handle (I am not musical at all) and changed all the notes to reflect what was needed to play the riff of *Smoke on the Water* by *Deep Purple*.

Additionally, I coded the LED lights on the Intelligent Hub to display the note or chord being played with each color. To do this, add a teal music block named **set instrument to** and choose the music tone you prefer. Next, add another teal music block named **play note**, and for the E note, choose 64 and change to a half-second note. Lastly, add a purple light block named **turn on** and edit the display to showcase the letter **E**:

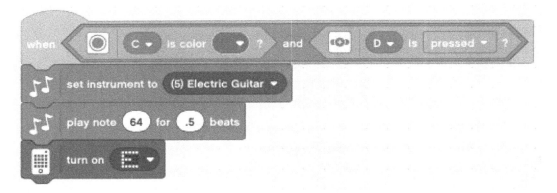

Figure 3.40 – The basic code of the guitar for each color

Once you have one color coded, you simply right-click on the selection of blocks and duplicate it four more times. For each one, you change the color of the color sensor and the note being played:

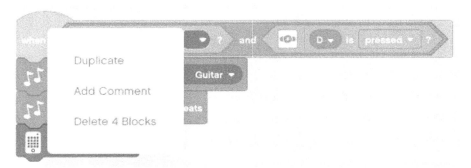

Figure 3.41 – Right-click so you don't have to code each block over and over

Here is how the code should look:

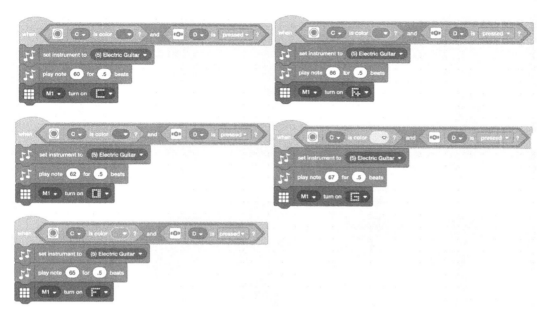

Figure 3.42 – The final code view

Now that the code is ready, let's play it!

Playing the guitar

The beauty of Bluetooth is that you can continuously tinker with your code while you play your guitar and do not need to be tethered to the computer. The music comes from the computer, so make sure your audio is turned up.

While playing the guitar, you can adjust how the music plays. I tried several options to find what I liked:

- Coding to simply see the color and to play without the force sensor
- Using the tap feature of the Intelligent Hub – when the brick is tapped, the note would play

Now that the guitar is built, coded, and playable, it's time to customize it to your own needs!

Making it your own

I can't wait to see how you design your guitar and how you play it. Here are a few other ideas that you could use to make your own custom guitar.

You could use the purple sound **play sound** block and import your sounds. Using the **record…** option, you could play power chords from the internet and record them to your Intelligent Hub. Using **edit sounds…** allows you to further customize the sound.

If you are talented, you could record yourself playing the actual chords/notes and pull that into the code:

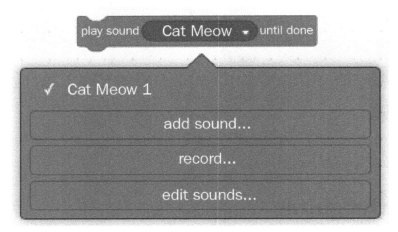

Figure 3.43 – Be creative and develop some unique sounds

You could also experiment using a block, such as the changing pitch block to change the pitch of a sound if the distance sensor is triggered (think of a whammy bar on an electric guitar) or if you tap the brick, or maybe add a motor to spin.

The possibilities are endless!

Figure 3.44 – Use distance sensors to tweak sounds like a whammy bar

Have fun. Rock out. Enjoy your build!

Summary

In summary, we explored how to make an instrument with a robotics kit. We remixed the idea of a robotics kit to make a musical instrument. There is great power in taking something we all know and love and trying to make a robotic version of it. We explored some new build techniques by using some of the basic elements that are found in the kit to create new ideas such as the guitar slider.

Finally, we explored coding by understanding the different ways we can take a simple code program and make it work to our liking.

In the next chapter, you will explore another aspect of life to see how you can adapt ideas learned from nature to build a robot with a killer instinct – by making a bird.

4
Building a Mechanical Bird

Biomimicry is the study and application of products, systems, mechanisms, and solutions to problems based on biological processes and functions found in nature. It is incredible what you can learn from plants and animals to find solutions to your own problems.

One of the many fascinating animals on our planet is the bird. It is a creature that has some features that are perfect for robot building and creating mechanisms. In this chapter, you are going to build a mechanical bird robot designed around the major features of this creature. You will be building wings and a body to look like a bird flying looking for prey.

Here is what your bird will look like by the end of this chapter:

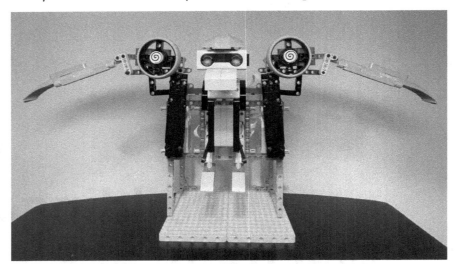

Figure 4.1 – This is what your build will look like by the end of this chapter

In this chapter, we will break down the building and programming as follows:

- Building the body frame of the mechanical bird
- Building the wings of the mechanical bird
- Building the head and torso
- Writing the code
- Activating the bird
- Making it your own

Technical requirements

For the building of the robot, all you will need is the **LEGO SPIKE Prime Kit**. For programming, you will need the LEGO SPIKE Prime app/software.

Access to the code can be found here: `https://github.com/PacktPublishing/Design-Innovative-Robots-with-LEGO-SPIKE-Prime/blob/main/Chapter%204%20-%20Bird.llsp`.

You can find the code in action video for this chapter here: `https://bit.ly/32is4IH`

Let's start building it!

Building the body frame of the mechanical bird

The beauty of this robotics kit is that you can easily get started with any type of build because of the new pieces that are included. You are going to use the two yellow *11x19* base plates to create a platform for the bird to be built upon:

1. Connect these two yellow base plates using two black connector pins along the long side of the base plates.

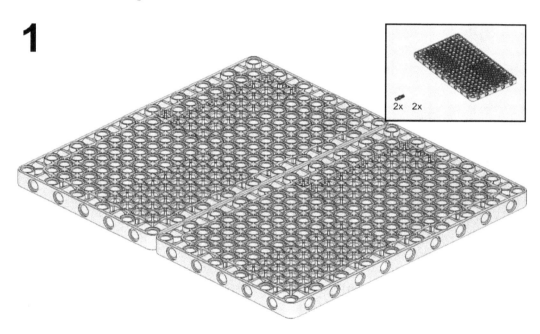

Figure 4.2 – Connecting two yellow base plates together

2. Add two azure *11x15* open frames to each end of the base you just assembled. Use two black connector pins on each end pin hole along with one gray perpendicular four-pin element to secure each open frame in place.

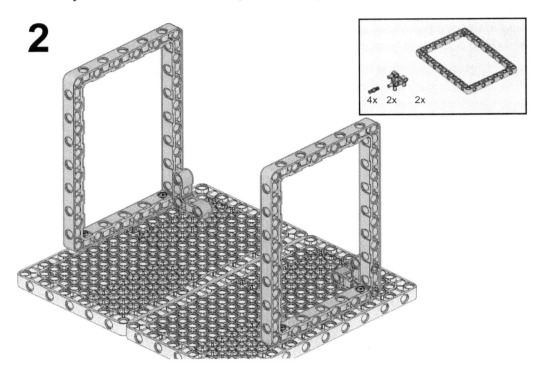

Figure 4.3 – Attaching two azure open frames to the yellow base plates

3. Now that the base frame is built to keep the bird in flight, it is time to build the frame of the body of the bird. Start with the Intelligent Hub. On each side of the Intelligent Hub, secure a purple *7x11* open frame using two gray bush connector pins. Additionally, add two black connector pins to the outside edges of the open frames, as shown in *Figure 4.4*:

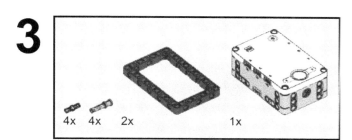

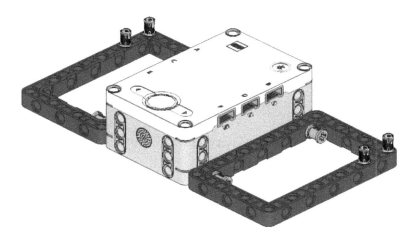

Figure 4.4 – Building the base frame of the bird's body

4. Using the black connector pins, attach the bird's body frame to the azure *11x15* open frames.

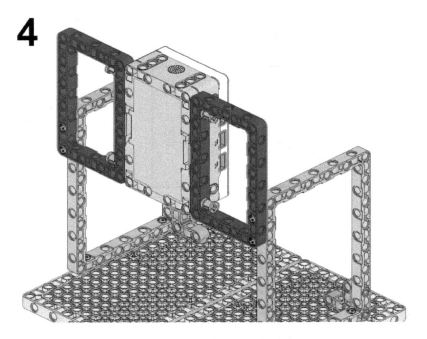

Figure 4.5 – Securing the bird frame to the azure open frames

5. It is now time to add some additional structural support. Starting on the outside top edge of each of the azure *7x11* open frames, attach a *3L* yellow beam using one black connector pin and one blue connector pin. Following this, attach an azure *3x5* L-shaped beam securing the Intelligent Hub to the azure *7x11* open frames again using one black connector pin and one blue connector pin on each element.

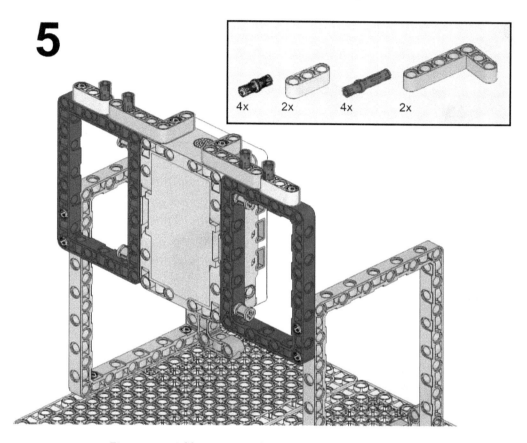

Figure 4.6 – Adding structural support to the Intelligent Hub

6. Add a purple *5L* beam to the top of the *3L* yellow beam and the *3x5* azure L-shaped beam using the blue connector pins still showing from the previous step. On top of the purple *5L* beams, attach a gray perpendicular four-pin element. Finally, connect a black *5x7* open frame to each of the gray perpendicular four-pin elements.

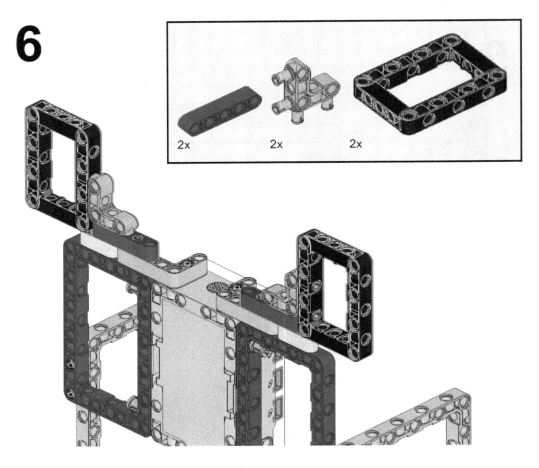

Figure 4.7 – Adding black open frames to the top of the build

The base frame for the bird is complete. You will continue to add to the body, but for now, it is time to transition to building the wing mechanisms for movement.

Building the wings of the mechanical bird

In order to have a quality-looking bird, you need wings that move and flap just like you see in the wild. These next steps will help create the mechanism for the wings to flap:

7. Inside each of the purple *7x11* open frames, insert a medium motor using two gray bush stop pins on the top and bottom of the side of each motor.

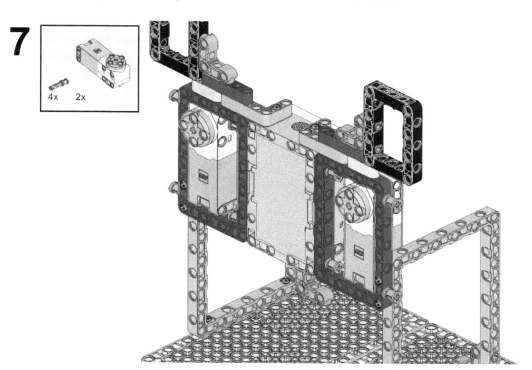

Figure 4.8 – Inserting the two medium motors

8. Just like all previous builds in this book, be sure your motors are in the 0 position by aligning the gray dots on the motor gear and frame. Once you have double-checked the alignment, attach a yellow *3L* beam to each of the motors using a tan axle pin and a black connector pin. Once the *3L* yellow beam is attached, add another black connector pin to each of the elements on the bottom pin hole.

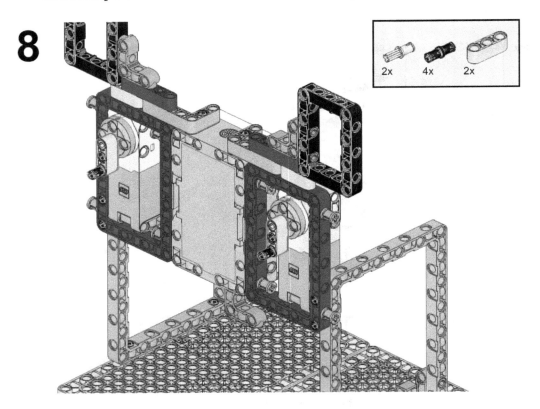

Figure 4.9 – Adding yellow 3L beams to medium motors

9. To add the wing structures, follow these steps:

A. Start by adding an azure *13L* beam to the black connector pin on each of the medium motors.

B. On the third pin hole from the other end on the azure *13L* beam, insert a *3L* pin connector with a *2L* axle using the axle side with a bush stop inserted on the *2L* axle side of the pin. Attach a purple *11L* beam to this pin on the third pin hole.

C. Add a black connector pin to the last pin hole of the azure *13L* beam. Attach a purple *11L* beam to the black connector pin using the end pin hole.

While the following figure shows these elements being held in place, the reality is that in this step, they will fall. You will secure them upright in the next step:

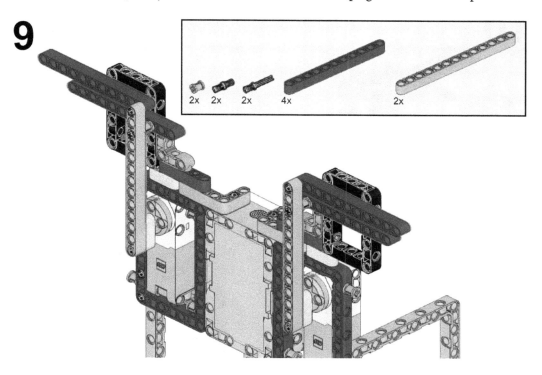

Figure 4.10 – Adding beginning wing structures

10. Use a black connector pin to secure the wings upright by adding this pin to the fourth pin hole of the bottom purple *11L* beam and connecting the pin to the black *5x7* open frame. Do this on both sides. On all four of the purple *11L* beams, insert a *3L* pin connector with a *2L* axle on each end with the axle side facing toward the front.

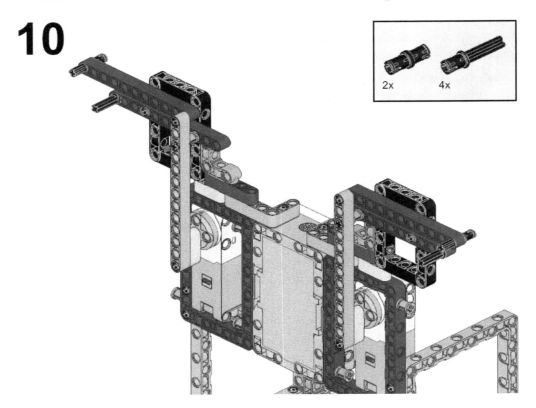

Figure 4.11 – Adding pins to extend the wings

11. You are going to continue to build out the wings in this next step. To begin, attach a blue connector pin to an azure *13L* beam. Leave the *2L* part of the blue connector out. Insert this azure *13L* beam on the black connector pin at the top end of the purple beam on the bird. Secure it in place using a yellow *2x4 L* beam connected to the blue connector pin and black connector pin. Do this for both wings.

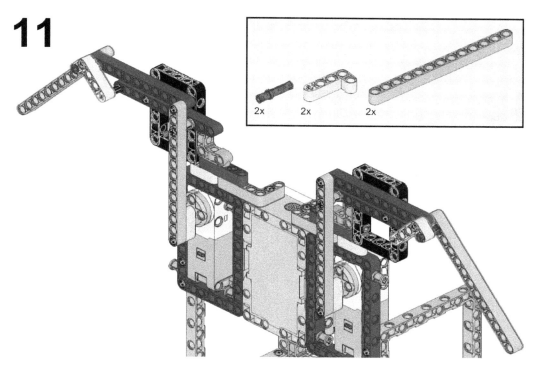

Figure 4.12 – Extending the wings

12. The steps to add the wings are as follows:

A. Attach three black connector pins to the azure *13L* beam and attach an azure *11x3* curved panel to this beam.

B. Connect an azure *7L* beam to the top of the *11x3* curved panel using a black connector pin. Be sure to leave two pin holes hanging over the end of the curved panel.

C. Snap an azure *8x3x2* wedge to the end and secure it using the two pin holes of the *7L* beam.

Do this for both wings:

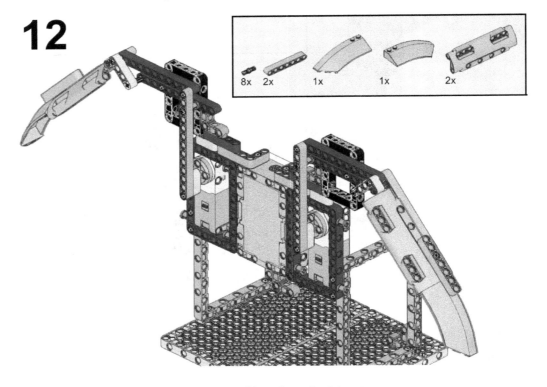

Figure 4.13 – Adding the ends of the wings

It is now time to begin working on the head and torso of the bird's build.

Building the head and torso

With the wings now ready, let's continue to build the body to cover up some of the wing mechanisms and bring the bird to life:

13. To begin this next section, attach the large motor to the top of the Intelligent Hub using four black connector pins. This motor will be used to rotate the head of the bird.

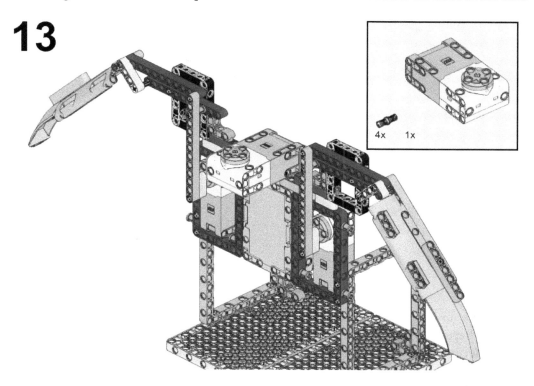

Figure 4.14 – Attaching the large motor to the Intelligent Hub

14. On the underside of the large motor, insert a gray *5L* axle beam on the motor.

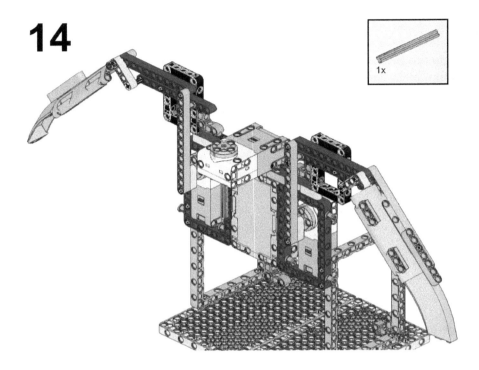

Figure 4.15 – Inserting an axle beam on the large motor

15. You are now going to build a submodel that will be used to build out the legs of the bird. Begin this submodel by inserting a tan connector pin on a yellow axle and pin connector #6. In *Figure 4.16*, you will see what the submodel will look like when complete in the upper right-hand corner:

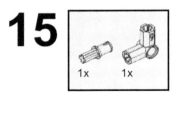

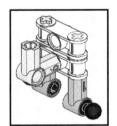

Figure 4.16 – Inserting a tan connector pin onto the yellow axle and pin connector

16. Stack two white axle and pin connector elements using two yellow *3L* axle beams. Connect these elements to the tan connector pin from the previous step into the bottom pin hole of the stack.

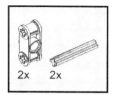

Figure 4.17 – Combining the white connector pin elements with the tan connector pin

17. On the underside of the yellow *3L* axle beams, insert a gray axle and pin connector *#1*. On the outside pin holes of the gray axle and pin connector *#1*, insert a black connector pin with a tow ball.

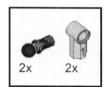

Figure 4.18 – Adding gray connector pin #1 and a black connector pin with a tow ball

18. Attach this submodel to the gray *5L* axle beam connected to the large motor using the yellow axle and pin connector *#6*.

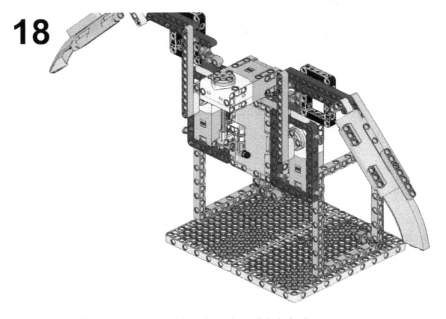

Figure 4.19 – Attaching the submodel to the large motor

19. It is now time to add the legs to the bird. You will be building another submodel. Begin by using two purple *2x4* bricks. Insert a red axle pin connector to each of them along with a white axle and pin connector *#4*.

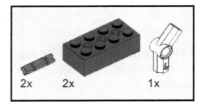
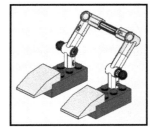

Figure 4.20 – Building the bird's feet

20. Insert a red axle connector pin on the top of each of the white axle and pin connectors #4. Attach a yellow axle and pin connector #6 to the top of each red connector pin. Secure them together using a black *4L* axle beam.

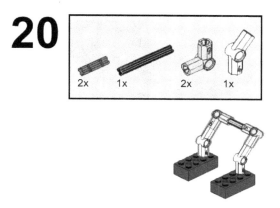

Figure 4.21 – Connecting the legs together

21. The last step of this submodel is to attach a yellow curved *3x2* slope to the end of each of the *2x4* purple bricks. Insert a black connector pin with a tow ball to each of the pin holes on the white axle and pin connector #4.

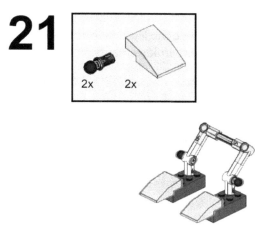

Figure 4.22 – Completing the legs

22. The legs are going to be added under the large motor and the submodel you built in *previous steps*. The following figure gives you a view of where it will be placed. At this point, the legs are not connected to the bird frame:

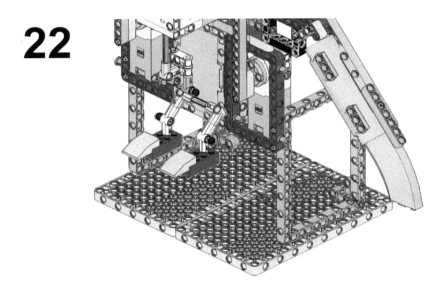

Figure 4.23 – Location of the submodel

23. Now that you have an idea of where the legs will be placed, you will connect them to the bird's torso using two black *1x6* link elements.

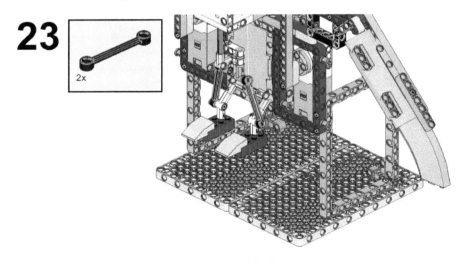

Figure 4.24 – Connecting the feet to the torso

24. Let's move to the head build now that the legs are complete. To begin, attach a black double bent beam to each side of the large motor using two black connector pins for each double bent beam.

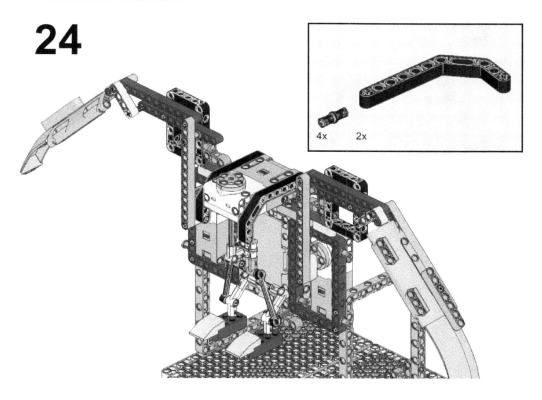

Figure 4.25 – Attaching black liftarms to the sides of the large motor

25. Connect a black *9L* beam to each of the black double bent beams using a black connector pin and a tan connector pin on each double bent beam.

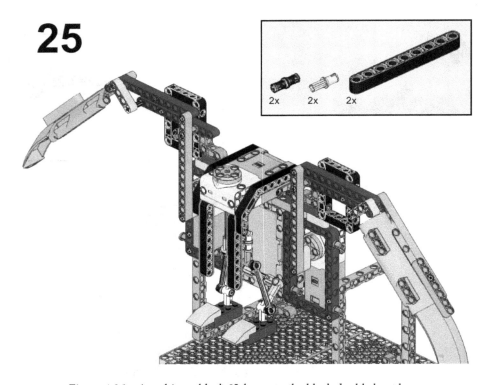

Figure 4.26 – Attaching a black 9L beam to the black double bent beam

26. When facing the front of the bird, insert two black connector pins seven pin holes apart. Insert the black connector pins in the third and sixth pin holes from the top of the black *9L* beam. Slide an azure *7L* beam onto the two black connector pins. Add two blue connector pins to each end of the azure *7L* beam.

26

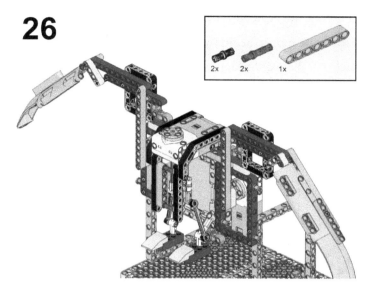

Figure 4.27 – Attaching the azure 7L beam to the black 9L beam

27. Slide two more azure *7L* beams onto the blue connector pins. Secure these three azure *7L* beams to the other black *9L* beam using two black connector pins.

27

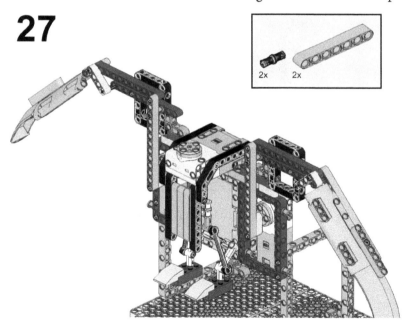

Figure 4.28 – Completing the body of the bird

28. To build the base for the beak and the head, follow these steps:

A. On top of the large motor, begin by adding two blue connector pins.

B. Insert a black *2x8* plate onto each of the blue connector pins.

C. Secure these plates together using a black round tile piece.

D. Insert a black biscuit element on the top of the blue connector pins vertically, as shown in the following figure.

E. Add two black connector pins to the black biscuit element.

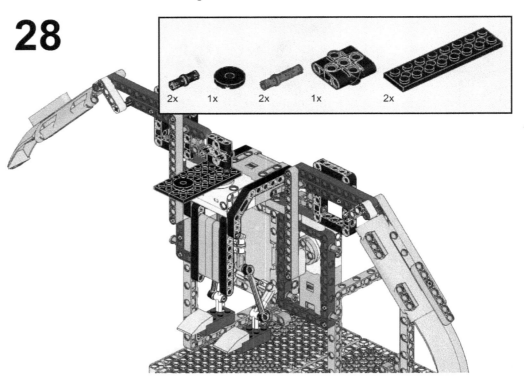

Figure 4.29 – Building the base for the beak and head

29. Insert two black *2x8* plates onto the black connector pins on the black biscuit element.

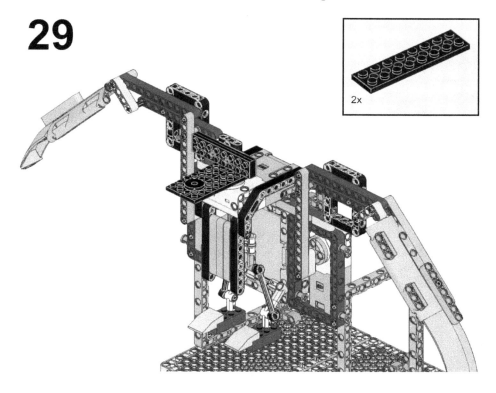

Figure 4.30 – Attaching two black 2x8 plates to the black biscuit element

30. Using the pin holes on the back of the ultrasonic sensor, add a black connector to each and secure them onto the black *2x8* plates.

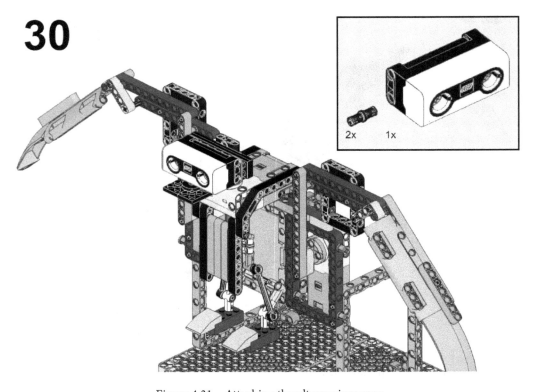

Figure 4.31 – Attaching the ultrasonic sensor

31. Here are the steps to build out the beak of the bird:

A. Add a yellow curved *3x2* slope to the end of each of the black *2x8* plates. Hold them together by adding a yellow *2x4* brick to the underside.

B. Add two black connector pins to the top of the black biscuit element.

C. Attach a black *6x2* slope to each of the black connector pins.

D. Attach an azure *8x6x2* windscreen element to the top of the black *6x2* slope.

31

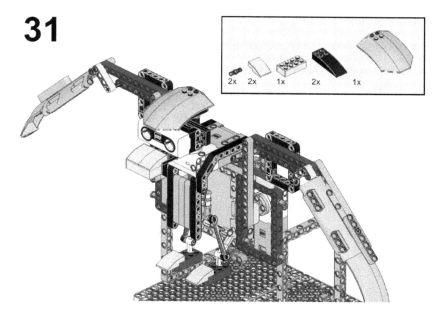

Figure 4.32 – Building out the head of the bird

32. Add two gray perpendicular pin connectors to the top purple beam on each wing. Insert a purple biscuit element on each of the gray perpendicular pin connectors.

32

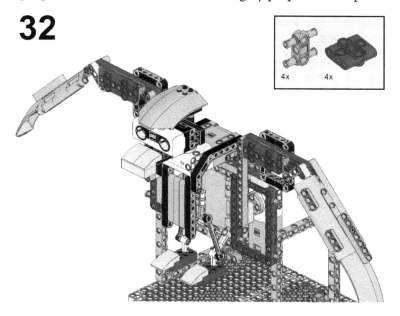

Figure 4.33 – Adding structure to the wings

33. This next set of steps will help with the look of the wings in the join areas:

A. Add two black connector pins to the inside purple biscuit element on each wing.

B. Attach a tire to these black connector pins.

C. Insert a red axle connector pin on the middle of each wheel.

D. Attach an azure *2x2* round plate to the red axle connector pin.

E. Attach a white spiral tile to the top of each azure *2x2* round plate.

F. Insert a black connector pin on pin holes four and six on the azure *13L* beam that is connected to each medium motor.

G. Attach a black biscuit element to these connector pins.

H. Attach two more black connector pins to the bottom of each of the black biscuit elements.

I. Attach a black *3x11x1* panel plate to these black connector pins.

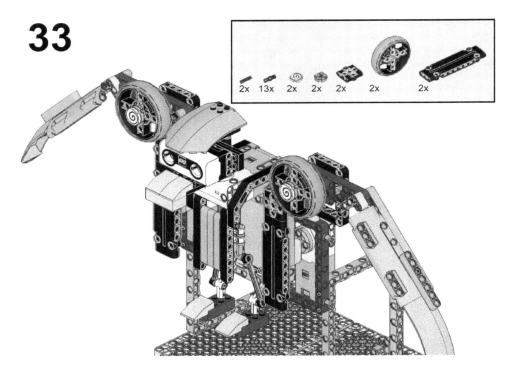

Figure 4.34 – Final decoration elements

Your bird is now complete. It is now time to write some code to bring the bird to life.

Writing the code

The code for the mechanical bird will use the ultrasonic sensor as both the eyes and to trigger movement. The goal is to mimic a bird in flight looking for prey, so we will work to ensure the legs move with the head gesture as well as the wing mechanics.

The ports

You will connect the ultrasonic sensor into port **B**. You will plug the large motor into port **F**. You will plug the two medium motors into ports **C** and **D**.

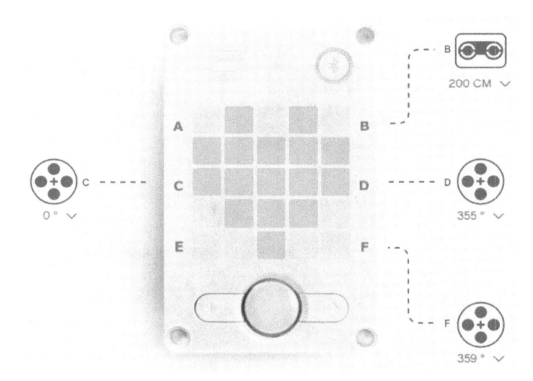

Figure 4.35 – Port view in the software

Calibrating the motors

This first code stack will make sure that the head and wing motors are all set to position 0 before being activated to ensure everything works smoothly and properly:

1. Under the **when program starts** block, insert a purple **light up** block that activates the lights on ultrasonic sensor **B** lighting up to **0 0**.

2. Add a pink movement block named **set movement speed to** and make it **15%**.

3. Add a blue **set speed to** motor block, set it to **10%**, and change it to motor **F**.

4. Add three blue **go shortest path to position 0** motor blocks. Change them to motors **C**, **D**, and **F**.

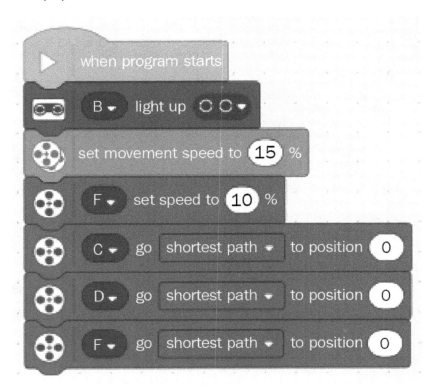

Figure 4.36 – Calibration of the motors code stack

Wing movement

You will create a new code stack to allow the wings to move when triggered by an object in view of less than 5 inches:

1. Add a yellow **event** block for the ultrasonic sensor that reads **when closer than**. Change it to port **B** and the distance to **5** inches.

2. Add a pink movement block named **set movement motors to** and set it to **C+D**.

3. Add another pink movement block named **move for 10 cm**. Change the setting to **5 rotations**.

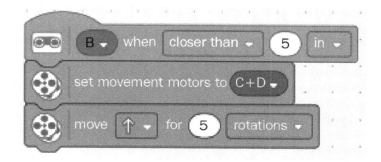

Figure 4.37 – Code stack for wing movement

Eyes and head movement

The final code stack will give the eyes the appearance of blinking, along with the head moving to scan for food:

1. Add a yellow **event** block for the ultrasonic sensor that reads when closer than. Change it to port **B** and the distance to **5** inches.

2. Add an orange control block named **wait** and change it to **.5 seconds**.

3. Add an orange control block named **repeat until**:

   A. In the space of this block, add a green operator block of **equal** comparison.

   B. On the left side of the comparison, insert a blue motor block with **C power**.

   C. On the right side of the comparison, type in a value of **0**.

4. Within the **repeat until** block, do the following:

   A. Add a purple light block for the ultrasonic sensor named **light up** and turn on all the lights.

B. Add a blue motor block named **go shortest path to position** for motor **F** and change it to **330**.

C. Add an orange control block named **wait** and change it to **.5 seconds**.

D. Add a blue motor block named **go shortest path to position** for motor **F** and change it to **30**.

E. Add a purple light block for the ultrasonic sensor named **light up** and turn off all the lights.

F. Add an orange control block named **wait** and change it to **.25 seconds**.

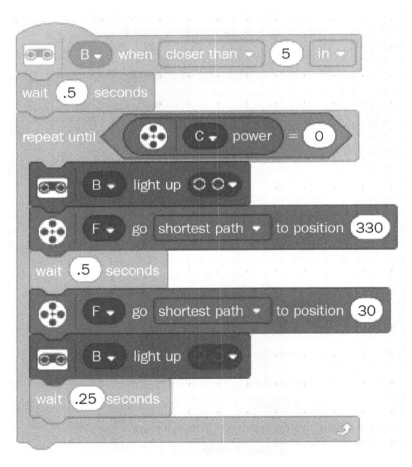

Figure 4.38 – Code stack for the eyes and head movement

Here is a look at the code altogether:

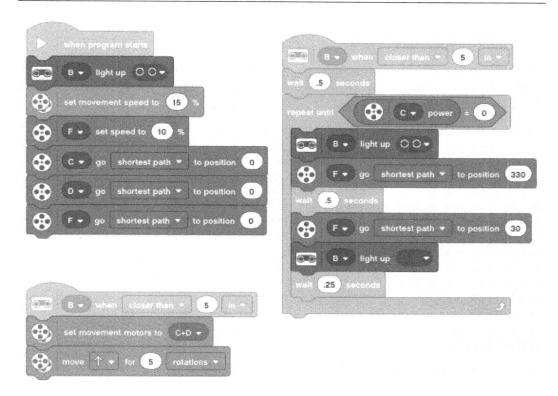

Figure 4.39 – Final look of the code

You have done it! Congratulations. Turn on the code and your bird and watch it come to life. Remember, this is just the start of the possibilities. What will you do to make it your own?

Activating the bird

This is just short sample code to get your bird started. You can continuously tinker with your code, changing various settings in the code as well as on the bird to get the bird working the way you choose.

While tinkering, here are some things I tried and found that I liked:

- Adjust the pins of the blue beams connected to the medium motors for different wing angles and movements.
- Adjust how the eyes blink, changing the light patterns.

Remember, the beauty of imagination and creativity is taking new learning and transforming it into whatever you desire. Let's explore some other ideas to spark your imagination.

Making it your own

Just like every build, the challenge here is to take this build and code and modify it to your needs. Here are some ideas to explore if you desire:

- Program the LED on the Intelligent Hub to mimic feather movement.
- Use the force sensor to trigger another behavior.
- Add sound effects.
- Build a model of food that can be clutched in the talons of the bird.

It is now up to you to transform your bird into the creature of your imagination. The possibilities are endless!

Summary

In summary, we explored how to make a mechanism powered by a motor to give the movement of wings of a bird in flight. You explored some new build techniques by using some of the basic elements that are found in the kit to help give a more realistic movement of the legs moving with the head while flying.

Finally, we explored the coding by using code stacks to make decisions. The eyes and head move if the power of motor C is not at 0. Once this motor stops, so do the head and eyes.

In the next chapter, you will explore building another robot by having a go at perhaps the most popular build idea – the sumobot.

5
Building a Sumobot

Sumobots and battlebots are both terms that you may have heard of before. In case you don't know what sumobots are, they are basically robots that are designed to battle one another. Basically, it is a sport where two robots battle in a head-to-head competition to outlast the other robot. Battle robots are a classic build challenge for any robot enthusiast. The popular TV show, *Battle Bots*, that has led to shows that are now on TV, Twitch, and other platforms, you can find robots battling in all types of arenas, with designs based on just about every conceivable idea you can imagine. These robots come in all shapes and sizes, depending on the rules of the competition. These robot challenges are quite popular in schools, after-school programs, and summer camps when it comes to **sumobots**. The robot you will build in this chapter will provide you with a solid foundation for being dominant in your next arena battle.

Here is a picture of what your sumobot will look like by the end of this chapter:

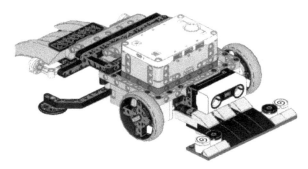

Figure 5.1 – Sumobot

In this chapter, you will break down the build and program as follows:

- Building the base
- Building the ultrasonic sensor and color sensor attachment
- Building the force sensor attachment
- Writing the code
- Making it your own

Technical requirements

For the building of the robot, all you will need is the **LEGO SPIKE Prime** kit. For programming, you will need the LEGO SPIKE Prime app/software.

Access to code can be found here: `https://github.com/PacktPublishing/Design-Innovative-Robots-with-LEGO-SPIKE-Prime/blob/main/Sumo.llsp`.

You can find the code in action video for this chapter here: `https://bit.ly/3oTqWmI`

Let's start building!

Building the base

The beauty of this robotics kit is that you can easily get started with any type of build because of the new pieces that are included. There are a lot of different options, depending on the strategy and size of your sumobot.

For this robot, the following strategies and constraints were used in the design:

- The body frame needs to be open for any user to remix, add to, and modify to their liking.
- The sumobot strategy is to identify the other robot and use sensors to push the other robot out of the arena using the little scoops at the front and back to slightly lift and push the robot.
- The robot needs to be low to the ground to keep the weight distribution and center of gravity low to avoid falling over.

To begin, you will add eight black connector pins to an azure *11x15* open frame:

1. Place four on each side, according to *Figure 5.2*. These are placed to allow the motors to be added:

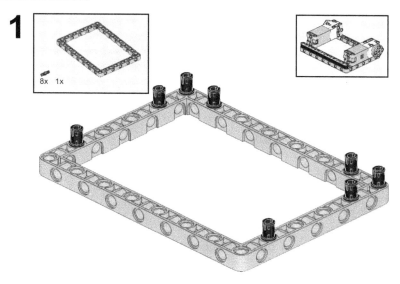

Figure 5.2 – The main frame of the robot body

2. Attach two medium motors to the black connector pins. Add a black *15L* beam across the backside of the open frame using two more black connector pins:

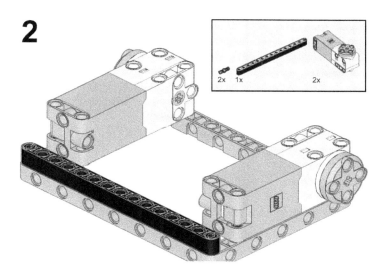

Figure 5.3 – Attaching the motors

3. Flip the frame over so the motors are on the underside:

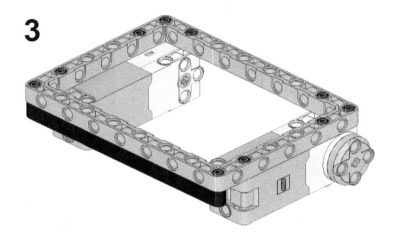

Figure 5.4 – Flipping the frame over

4. Insert four black connector pins into the corners of the medium motors:

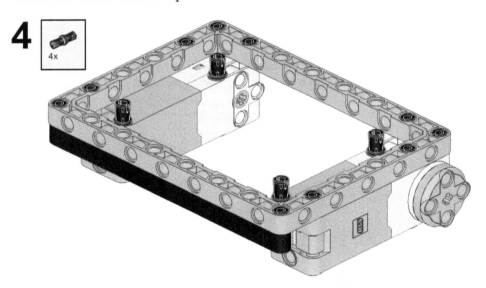

Figure 5.5 – Inserting the black connector pins into the medium motors

5. Add a purple *7x11* open frame to the black connector pins. Add four more black connector pins to the purple *7x11* open frame. These will be used to attach the Intelligent Hub:

5

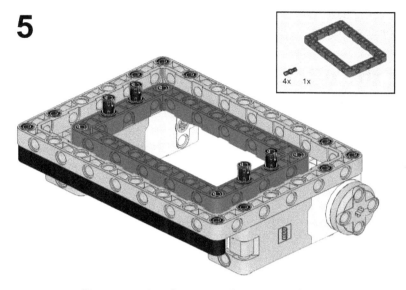

Figure 5.6 – Attaching a purple 7x11 open frame

6. Attach the Intelligent Hub to the purple *7x11* open frame using the four black connector pins:

6

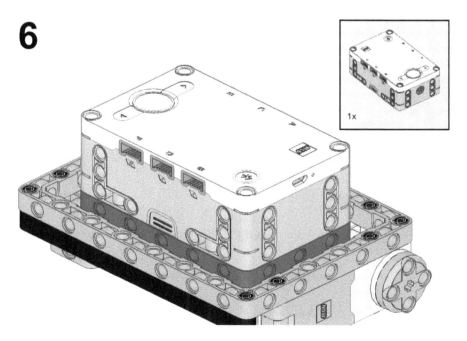

Figure 5.7 – Attaching the Intelligent Hub

The simple but secure base has now been completed. It is time to build and attach the ultrasonic sensor attachment.

Building the ultrasonic sensor attachment

You are now going to add the ultrasonic sensor so that during competition, your robot can detect the opponent:

7.   Add two black connector pins to the azure *11x15* open frame where the motors are positioned and then add an azure *7L* beam to these pins. Next, add two more black connector pins to the end pin holes of this beam:

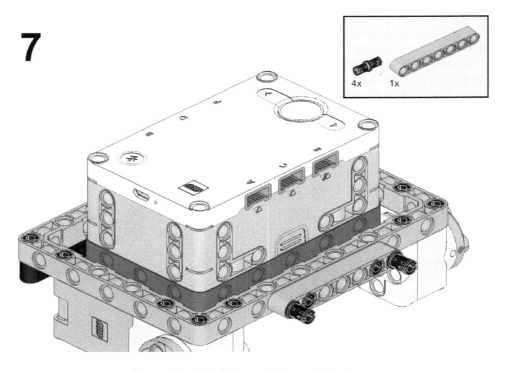

Figure 5.8 – Attaching the 7L beam to the frame

8. Connect the ultrasonic sensor to the robot by using the two black connector pins:

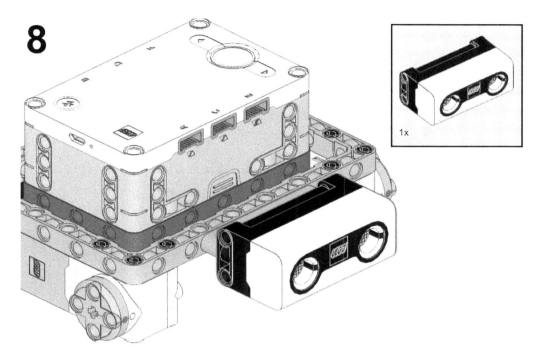

Figure 5.9 – Connecting the ultrasonic sensor

9. You are now going to build a sub-model that will eventually connect underneath the ultrasonic sensor using the following steps:

    A. Place two purple *2x16* plates side by side.

    B. Secure them together using four white *2x2* round bricks.

C. Add four yellow *1x2* slopes on the ends to serve as a bumper to lift up your robot opponent in battle:

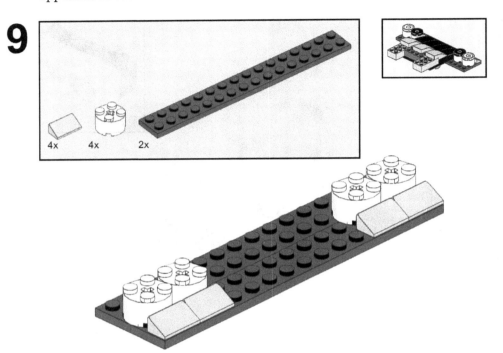

Figure 5.10 – Building the plow for the front

10. It is time to add another layer to the plow using the following steps:

A. Attach four of the black *2x6* slopes to the purple plates. This strengthens the scoop and helps hold everything together.

B. Add three yellow *2x3* slopes to the tops of the black *2x6* slopes:

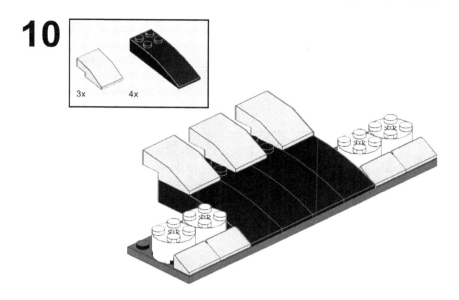

Figure 5.11 – Adding another layer to the plow

11. Add three *2x4* bricks to the underside of the yellow *2x3* bricks. The outside yellow *2x4* bricks are parallel with the black curved panels. The red *2x4* brick is perpendicular to the black curved panel:

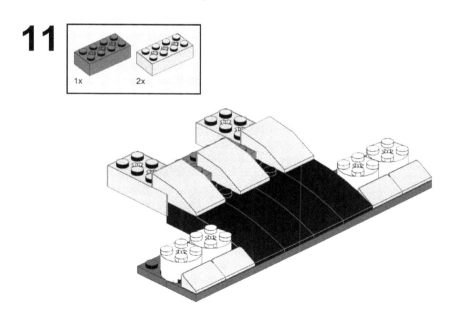

Figure 5.12 – Attaching 2x4 bricks to the underside of the yellow slopes

12. The following set of steps will focus on the underside of the robot. Begin by flipping the build over:

A. Add a black *2x8* plate to the yellow *2x4* bricks.

B. Hold things in place by adding four azure *2x2* round plates across the black *2x8* plate and the black *2x6* slopes:

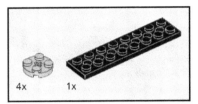

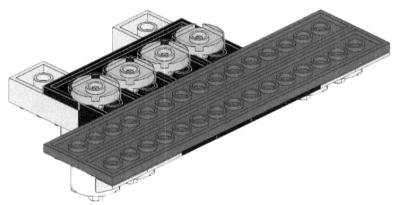

Figure 5.13 – Securing the parts of the plow

13. Flip this model back over and add two white spiral tiles and two black tiles for some decorative touches:

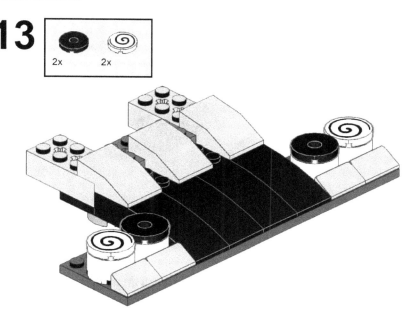

Figure 5.14 – Adding some design to the plow

14. Before you build the next part of the plow that houses the color sensor, here is how it should look now:

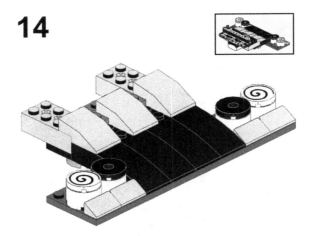

Figure 5.15 – A view of the front plow

15. You are going to build another sub-model that will attach to the plow you just built. This sub-model houses the color sensor. The steps are as follows:

A. Add two black connector pins to the two black biscuit elements. Add one connector pin to the outside top pin hole and another to the side pin hole.

B. Add two black connector pins to the color sensor:

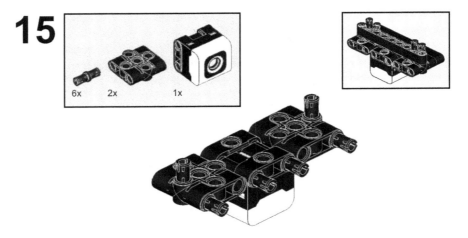

Figure 5.16 – Adding pins to the color sensor model

16. Hold the two biscuit elements and color sensor together using two black *9L* beams by attaching them to the black connector pins. On the exposed pin hole of the black biscuit elements, insert a black pin and an axle connector. These will not stay secure now, but you will take care of that in the next step:

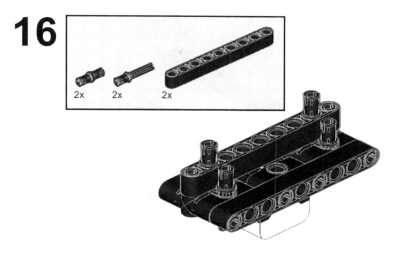

Figure 5.17 – Securing the parts together

17. Using the black pin and axle connector pins from the previous step, attach this color sensor sub-model to the plow by adding it to the axle holes in the yellow *2x4* bricks:

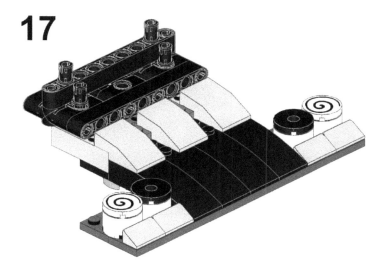

Figure 5.18 – Combining the plow and color sensor sub-models

18. This whole sub-model now attaches to the body of the robot. It will join using the black connector pins to the underside of the azure *11x15* open frame between the motors:

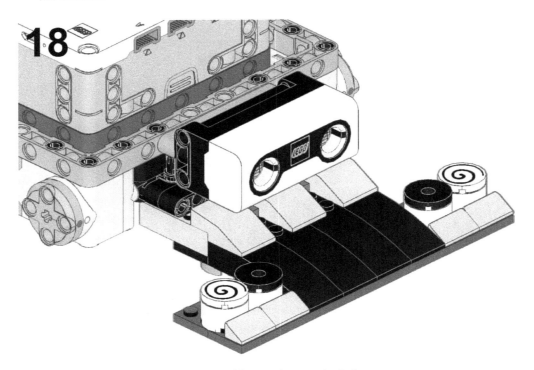

Figure 5.19 – Adding to the main body frame

Now that the front is complete, it is time to begin to build the back attachment using the force sensor.

Building the force sensor attachment

Before we add the force sensor, the robot needs to be lifted up with wheels and the plastic caster ball element using the following steps:

19. To begin this process, connect a gray *3x5* H-shaped beam to the azure caster ball frame using two black connector pins:

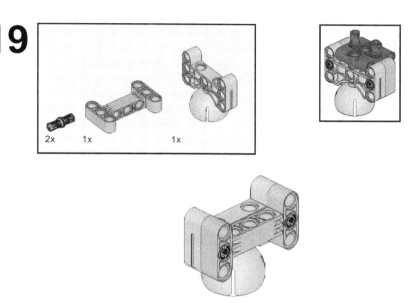

Figure 5.20 – Attaching the H frame beam to the caster ball element

20. On the top of these two elements, add a purple biscuit element using two blue connector pins:

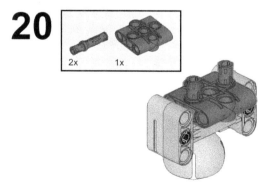

Figure 5.21 – Add a biscuit element to the top

21. This sub-model will attach to the underside of the robot body to the purple *7x11* open frame. This element is placed opposite to the color sensor:

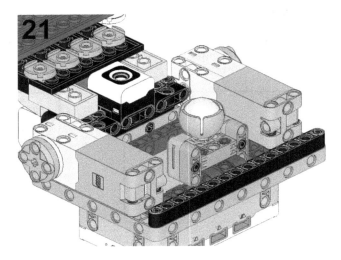

Figure 5.22 – Attaching the caster wheel to the underside of the robot

22. Next, turn the sumobot back to the original position. You will add a wheel to each of the medium motors. Attach a wheel using two black connector pins and one yellow *3L* axle:

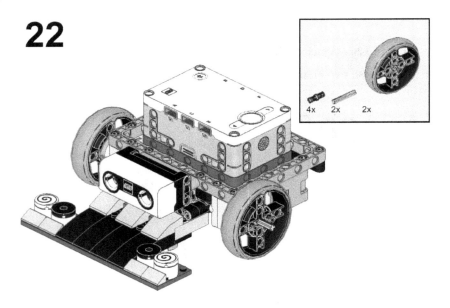

Figure 5.23 – Adding the wheels

Your robot should not be balanced and ready for action. Before you move into action, you need to add the force sensor attachment using the following steps:

23. Begin by connecting the force sensor to the backside of the robot body using two gray perpendicular connector elements. Add these elements to the sensor first and then connect collectively to the black *15L* beam on the robot body frame:

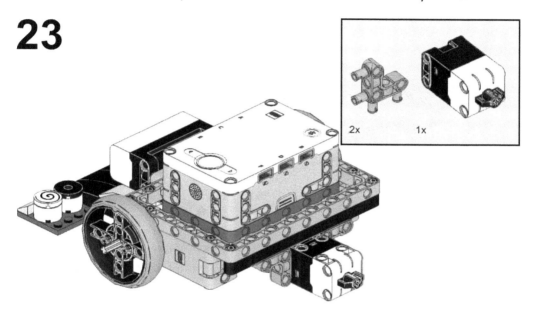

Figure 5.24 – Attaching the force sensor to the body

24. It is time to build out the bumper for the sensor so that your robot can feel the impact from any angle of attack:

    A. Start with a black *15L* beam.

    B. Insert a black pin and axle connector pin to the middle pin hole with the axle element being exposed.

    C. Insert two black connector pins on either side of this pin.

    D. Attach a purple *5L* beam to the pins.

E. Insert this build piece into the force sensor:

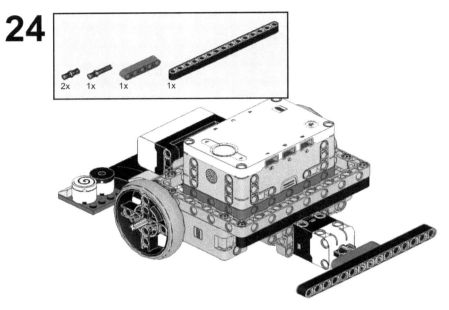

Figure 5.25 – Building out the force sensor bumper

25. Insert two black connector pins and two blue connector pins into the black *15L* beam, as shown in *Figure 5.26*:

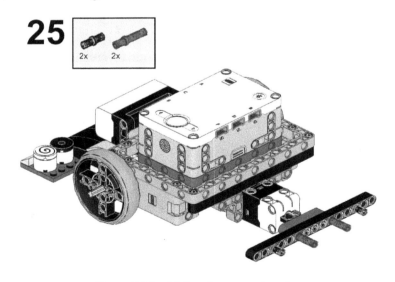

Figure 5.26 – Adding the connector pins

26. To help the bumper work to activate the force sensor, you need to add some additional elements:

A. Begin with a yellow axle and pin connector #6, and add a red axle connector pin to one end.

B. Attach a black axle and pin connector #2 to the red axle connector pin, and add another red axle connector to the other end.

C. Attach another yellow axle and pin connector #6 facing outward. Add a yellow 3L axle to the other end.

D. Slide a gray bush on the yellow 3L axle and a white wheel hub:

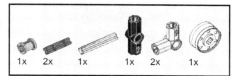

Figure 5.27 – The force sensor bumper extension

27. You will do this same build but mirror it for the other side:

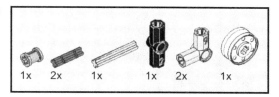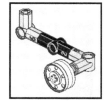

Figure 5.28 – Mirroring the build for the other side

28. Attach these two builds to the force sensor bumper using the pins on the black *15L* beam:

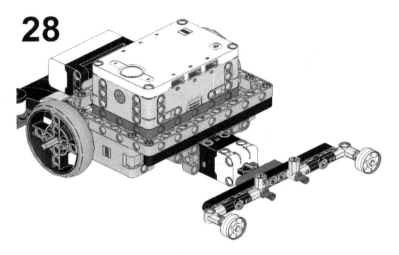

Figure 5.29 – Adding builds to the bumper

29. Add an azure *7L* beam to the blue pins available on the front of the force sensor bumper. Add two black connector pins to this beam:

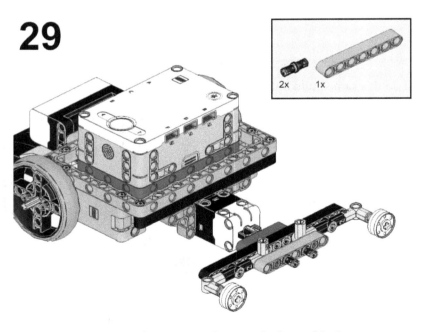

Figure 5.30 – Attaching an azure beam to the front of the bumper

30. It is time to add the plow to this sensor attachment using the following steps:

A. Begin by connecting the azure *8x6x2* windscreen element to the black connector pins on the front of the bumper.

B. Attach a black *2x8* plate to the top of the windscreen element.

C. Add an azure *8x3x2* wedge to both sides of the windscreen element.

D. Attach two black *2x8* plates to the bottom of the windscreen element to hold everything in place. Add one to the front and another to the back of these elements:

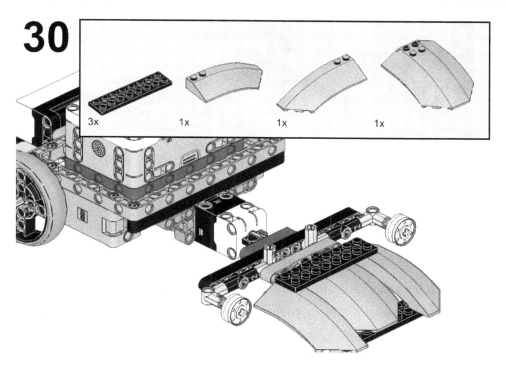

Figure 5.31 – Building the plow for this sensor

31. On top of the yellow axle and pin connectors #6, add a tan pin and axle connector. Attach an azure *7L* beam to both pins:

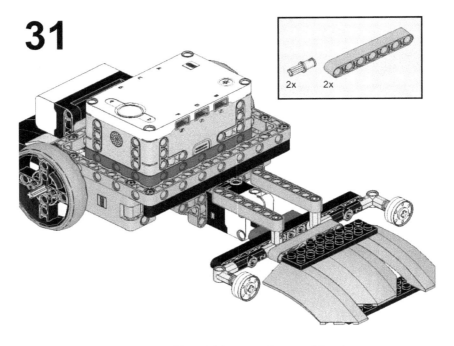

Figure 5.32 – Adding the beam to the top of the plow

32. It is time for the final steps to ensure the bumper stays in place:

A. On both the azure *7L* beams, add a black connector pin to the end pin hole.

B. Attach a black biscuit element to these pins.

C. Add a purple *5L* beam across to connect the azure *7L* beams.

D. Do this same process again but on the third pin hole from the end.

E. In the middle pin hole on both purple *5L* beams, add a black connector pin.

F. On the underside of the azure *7L* beams, insert a gray perpendicular pin connector into both sides.

G. Finally, add a yellow *3L* beam to the bottom of both gray perpendicular pin connectors:

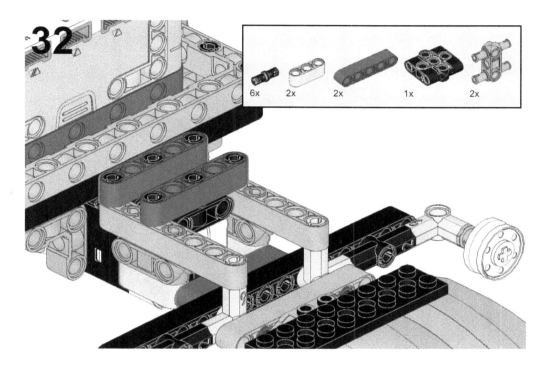

Figure 5.33 – Adding the frame to keep the bumper in place

33. You are closing in on the final steps. The next part is to help your robot roll away from an attack as well as to roll an opponent into your bumper for your own attack.

Begin by adding a blue connector pin to a black double-bent lift-arm beam. Do this two times – once for the right side and once for the left side of your robot. Do not connect to the robot yet. The figure is to give you a sense of location:

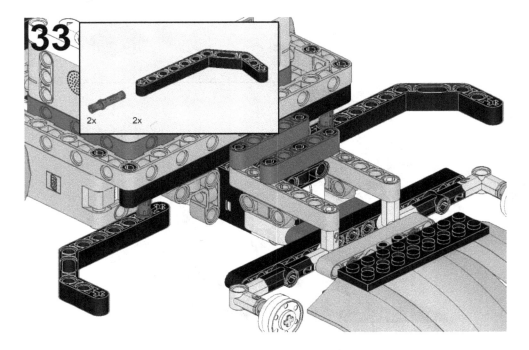

Figure 5.34 – Building roll-away attachments

34. Attach an azure *2L* beam with a pin and axle holes to the blue connector pin. Insert a black axle and pin connector through the axle hole of the azure *2L* beam and black double-bent lift-arm beam.

Once you have done this, then attach it to each side of the robot body on the black *15L* beam of the main robot frame:

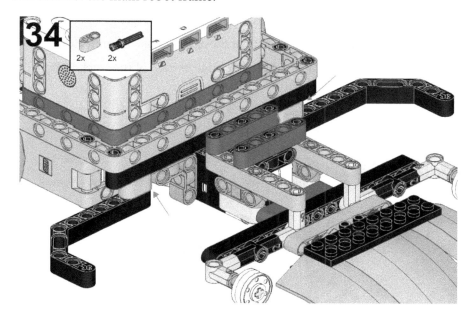

Figure 5.35 – Attaching the roll-away attachment to the robot

35. At each of the black double-bent lift-arms, add a tan connector pin to the second pin hole. Attach a pulley wheel hub to the tan connector pin:

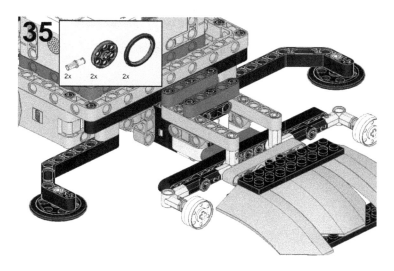

Figure 5.36 – Attaching the roll-away wheels

36. The last step is to add a few elements to round out the body of the robot for looks. The steps are as follows:

A. Add two black connector pins on each side of the Intelligent Hub.

B. Insert a black *5x7* open frame to each of these sets of black connector pins.

C. Add a black connector pin to the outside edge of each of the black *5x7* open frames:

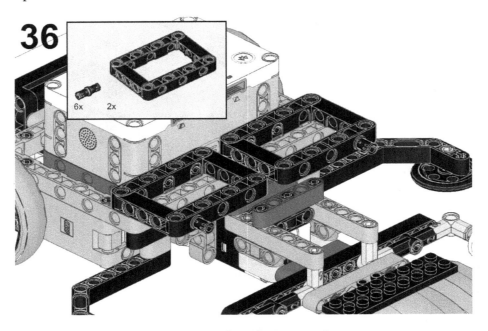

Figure 5.37 – Attaching the 5x7 open frames

37. Attach an azure curved panel to the Intelligent Hub using two black connector pins. These pieces will sit on top of the black *5x7* open frames.

Add a black *15L* beam across the two black *5x7* open frames. Attach another azure curved panel to this black *15L* beam using two black connector pins:

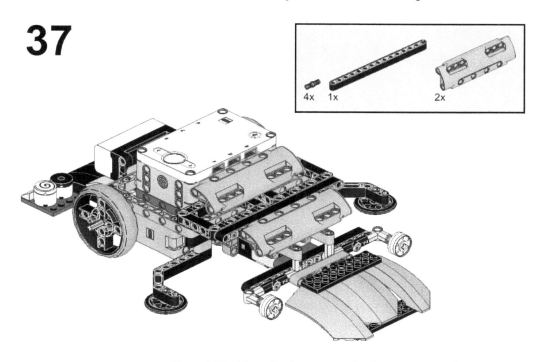

Figure 5.38 – The robot is now complete!

When it comes to organizing the wires, the key is to keep them away from the wheels and sensors. Also, it is important to have them away from the sides where another robot can get entangled in them. There is no perfect way to work the wires, but here is one way you can do it:

Figure 5.39 – The robot with wires on top

That's it! You have successfully built the sumobot. The fun now shifts to the coding as you design a program to keep the robot in the arena and looking to strike your opponent.

Writing the code

The code you are going to create is one where the robot will be able to do the following:

- Stay in the arena by stopping and backing up when the white rim of the arena is detected.

- Speed up to push an opponent out of the arena when the ultrasonic sensor detects an object up close.

- Speed up to push an opponent out of the arena when the force sensor detects touch from behind.

In all, there are six code stacks to make this happen. Let's begin.

The ports

You will connect the medium motors to ports **C** and **F**, the color sensor to port **A**, the ultrasonic sensor to port **D**, and the force sensor to port **E**:

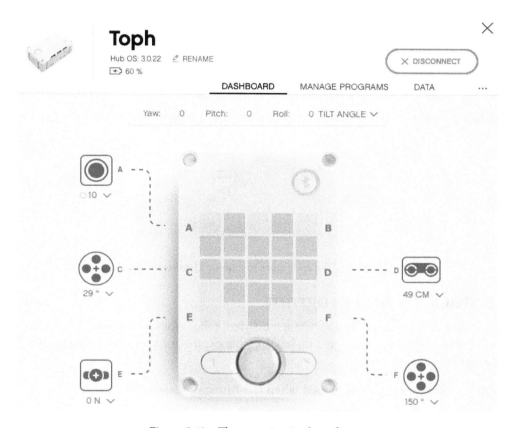

Figure 5.40 – The port view in the software

Before you code, you will need to add a **More Movement** extension block to your coding options. At the bottom left of your coding canvas, there is an extension block icon. Click on this icon, click on **Show block extensions** and select **More Movement**:

Figure 5.41 – Show block extensions

Now that your programming canvas is set up and ready, it is time to begin to build the first code stack that will activate the motors.

Code stack 1 – motors

This short code stack is just identifying which motors are for movement. It is kept open in case you want to modify and add more to it on your own. Let's begin adding the coding blocks:

1. Add a yellow events block named **when program starts**.

2. Add a pink movement block named **set movement motors to** and choose **C** and **F**:

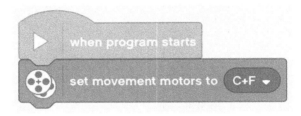

Figure 5.42 – The code stack for motor identification

Code stack 2 – end program

This code stack is designed to be able to turn off your robot when you need it to stop running. This is helpful if robots get tangled up, time expires, or you win! The steps for end program are as follows:

1. Add a yellow events block named **when left Button pressed**.

2. Add an orange control block named **stop all**:

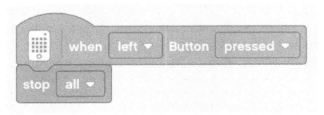

Figure 5.43 – The code stack for end program

Code stack 3 – main program

This is the brains of the operation. This code stack will help the robot make decisions based on a series of events and actions. Again, this is designed in a way that you can always create your own sumobot and adjust its behaviors.

The following steps will be the main program:

1. Add a yellow events block named **when program starts**.

2. Add a purple light block named **turn on for 2 seconds** and change the graphic to number **3** and the time to **1** second.

3. Add a purple light block named **turn on for 2 seconds** and change the graphic to number **2** and the time to **1** second.

4. Add a purple light block named **turn on for 2 seconds** and change the graphic to number **1** and the time to **1** second.

5. Go to the dark orange variable block section and click **Make a Variable**, and name it **action**.

6. Add a dark orange variable block named **set action to 0** and change it to **1**:

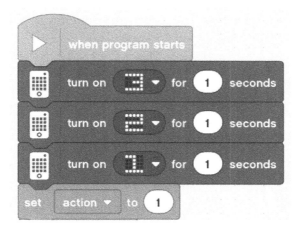

Figure 5.44 – The first five code blocks for this code stack

7. Add an orange control block named **forever**.

8. Inside the **forever** block, add an orange control block named **if then**:

   A. Add a green operator block with an equal sign. Make the left side the orange variable block named **action** and the right side the numerator **1**.

B. After the **then** statement, add a pink more movement block named **start moving straight at 50%** and change the percentage to **-25%**:

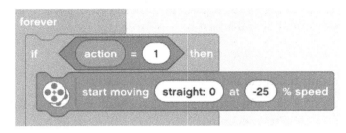

Figure 5.45 – Action variable when 1

9. Add another orange control block named **if then**:

A. Add a green operator block with an equal sign. Make the left side the orange variable block named **action** and the right side the numerator **2**.

B. Add a pink more movement block named **move straight for 10 cm at 50% speed** and change to **2 rotations at 75% speed**.

C. Add a pink more movement block named **move straight for 10 cm at 50% speed** and change to a green operator block named **pick random** for the direction, with a range of **-100 to 100 for 2 seconds at 75% speed**.

D. Add a dark orange variable block named **set action to** and choose numerator **1**:

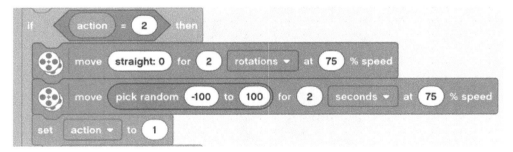

Figure 5.46 – Action variable when 2

10. Add another orange control block named **if then**:

A. Add a green operator block with an equal sign. Make the left side the orange variable block named **action** and the right side the numerator **3**.

B. Add a purple sound block named **play sound Cat Meow until done**.

C. Add a pink movement block named **start moving straight at 50%** and change to **-75%**.

D. Add an orange control block named **wait until** and add a blue operator block named **ultrasonic D is closer than 15%** and change to **farther than 8 in**.

E. Add a dark orange variable block named **set action to** and choose numerator **1**:

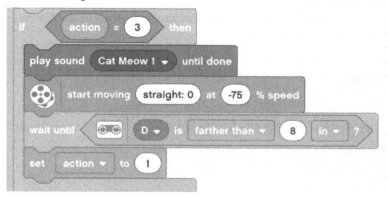

Figure 5.47 – Action variable when 3

11. Add another orange control block named **if then**:

A. Add a green operator block with an equal sign. Make one side the orange variable block named **action** and the other side the numerator **4**.

B. Add a pink movement block named **stop moving**.

C. Add a pink more movement block named **move straight for 10 cm at 50% speed** and change to **1** rotation at **75%** speed.

D. Add a dark orange variable block named **set action to** and choose numerator **1**:

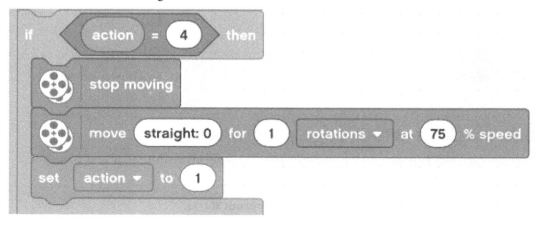

Figure 5.48 – Action variable when 4

12. Add another orange control block named **if then**:

A. Add a green operator block with an equal sign. Make the left side the orange variable block named **action** and the right side the numerator **5**.

B. Add a pink movement block named **start moving straight at 50%** and change to **75%** speed.

C. Add an orange control block named **wait until** and add a blue force sensor block when sensor is released.

D. Add a pink movement block named **stop moving**.

E. Add a dark orange variable block named **set action to** and choose numerator **1**:

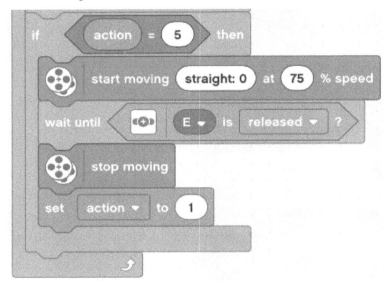

Figure 5.49 – Action variable when 5

Now that you have programmed the various actions of the robots, you now need to write some code to enable these actions to occur.

Code stack 4 – white edge program

This code stack will help determine whether your robot sees the white edge and moves into action accordingly. The steps are as follows:

1. Add a yellow events block named **when program starts**.

2. Add an orange control block name **forever**:

A. Add an orange control block named **if then**:

   i.   Change the **if** statement to a blue sensor block named **color reflection < 50%** and change to **> 90%**.

   ii.   Add a dark orange variable block named **set action to** and choose numerator **2** for the **then** statement:

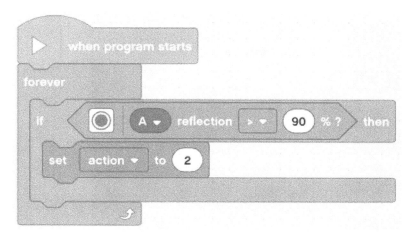

Figure 5.50 – The code stack for when the robot sees white

Your color sensor is now programmed and now you need to program the ultrasonic sensor.

Code stack 5 – ultrasonic program

This code stack will help the robot decide what to do when an object is detected. The steps are as follows:

1.   Add a yellow events block named **when program starts**.

2.   Add an orange control block name **forever**:

   A. Add an orange control block named **if then**:

      i.   Change the **if** statement to a blue sensor block named **ultrasonic is closer than 15%** and change to **6** inches.

      ii.   Add a dark orange variable block named **set action to** and choose numerator **3** for the **then** statement.

   B. Add an orange control block named **if then**:

      i.   Add a green operator block named **and** for the **if** statement.

      ii.   Change one side of the **and** statement to a blue sensor block named **ultrasonic is closer than 15%** and change to **6** inches.

iii. Change the other side to a blue sensor block named **color reflection < 50%** and change to **> 90%**.

iv. Add a dark orange variable block named **set action to** and choose numerator **4** for the **then** statement:

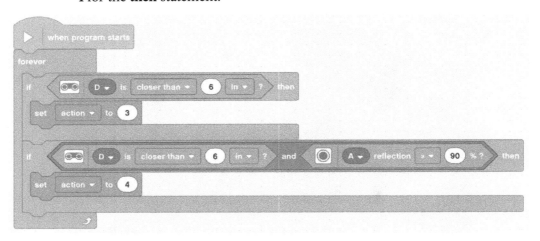

Figure 5.51 – The code stack for when the robot detects an object with the ultrasonic sensor

You are now down to one sensor left to program. It is time to write some code for the force sensor.

Code stack 6 – force sensor program

Just like you programmed earlier for the ultrasonic sensor, you need to write code for the force sensor to detect the impact when your robot collides with your opponent.

The following steps will code your robot to detect collisions and move into attack mode:

1. Add a yellow events block named **when program starts**.

2. Add an orange control block named **forever**:

    A. Add an orange control block named **if then**:

    i. Change the **if** statement to a blue sensor block named **force sensor is pressed**.

    ii. Add a dark orange variable block named **set action to** and choose numerator **5** for the **then** statement.

B. Add an orange control block named **if then**:

   i.   Add a green operator block named **and** for the **if** statement.

   ii.   Change one side of the **and** statement to a blue sensor block named **force sensor is pressed**.

   iii. Change the other side to a blue sensor block named **color reflection < 50%** and change to **> 90%**.

   iv. Add a dark orange variable block named **set action to** and choose numerator **4** for the **then** statement:

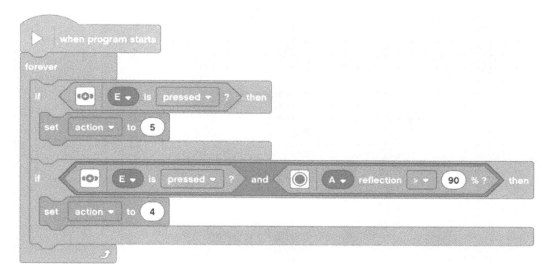

Figure 5.52 – The code stack for when the robot detects touch with the force sensor

Your code is now complete and should look something like *Figure 5.53(A)* and *Figure 5.53(B)* when all done:

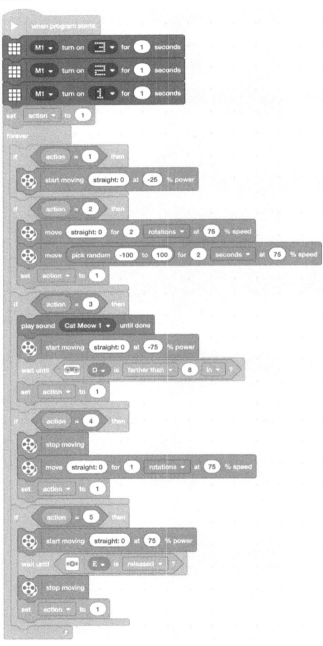

Figure 5.53(A) – The complete view of code

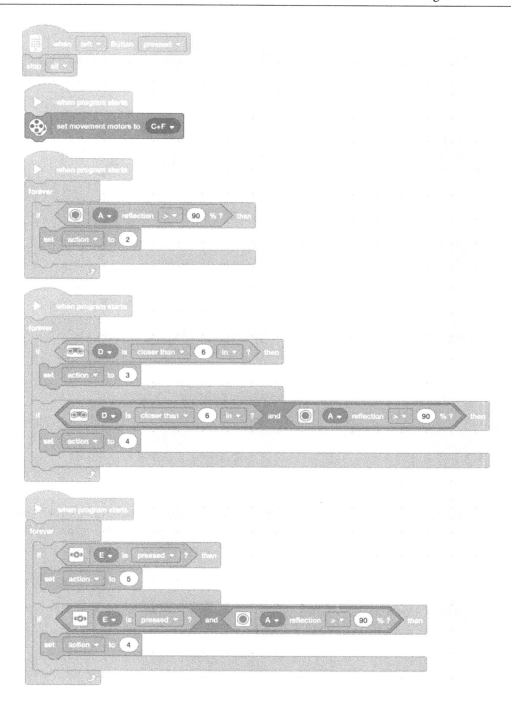

Figure 5.53(B) – The complete view of code

Go ahead and compete and see how your robot does. Remember, you have the power, imagination, and intelligence to make the robot better to your liking.

Making it your own

Just like every build so far, the challenge here is to take this build and code and modify it to your needs. Here are some ideas to explore if you desire:

- Implement another battle technique using the large motor.

- Gain more power by gearing your motors and wheels.

- Redesign the plow of the sensors to something else.

- Rewrite the code to have your robot behave in new ways.

- If you need help making a sumobot arena, then this website is a great place to get started: `http://www.robotroom.com/SumoCircleMini.html`.

- If you want to see examples of sumobot battles, then check out my EV3 camp website where you can see videos. Keep in mind these use LEGO Mindstorms EV3 as SPIKE Prime was not available in previous years: `- https://sites.google.com/view/lego-ev3-camp-2019/sumobots`.

- If you need a portable sumobot arena, this is one solution: `https://www.youtube.com/watch?v=QeKa_r-OrK0&feature=youtu.be`.

As you explore and try new methods, be sure to document and share so others can learn from your genius.

Summary

In this chapter, you explored how to build a robot that can compete in sumo competitions. You built out some attachments to coordinate with sensors so that your robot can make decisions and respond to an opponent in the field. The robot capitalized on a few standard strategies involving being low to the ground, allowing it to push an opponent by lifting them up.

Finally, we explored the coding by using code stacks to make decisions using a variable named `action`. This allows you to make a more precise decision-making robot.

In the next chapter, you will explore another robot that will succeed in another type of competition built for speed – the dragster!

6
Building a Dragster

Dragsters that race and are designed for speed are a classic build and one that every builder creates at some point in their lives. I have operated a **LEGO MINDSTORMS** robotics summer camp for several years and the dragster challenge is a favorite every year that the kids always want to compete in. Whether designing for yourself or competing against others, racing is a blast. In this chapter, you will build a dragster using the **SPIKE Prime** kit to see how it turns out in terms of speed and design.

Here is a photo of what your dragster will look like by the end of this chapter:

Figure 6.1 – Dragster

In this chapter, you will build and program as follows:

- Building the motor frame
- Building the gear system and back wheels
- Building the body of the dragster
- Adding the body design
- Coding the dragster

Technical requirements

For the building of the dragster, all you will need is the SPIKE Prime kit. For the programming, you will need the LEGO MINDSTORMS app/software.

Here is the link to the code: `https://github.com/PacktPublishing/Design-Innovative-Robots-with-LEGO-SPIKE-Prime/blob/main/Dragster.llsp`.

You can find the code in action video for this chapter here: `https://bit.ly/3FJPx41`

Building the dragster

Before we get into the building of this dragster, let's explore the strategy being used for this dragster. The strategy for drag racing is all about speed. How fast can you get from the starting line to the finish line? That is it! The design of this dragster is based on the following strategies:

- Low to the ground to eliminate drag.
- Minimal touch points – the less friction the better.
- Gearing up to max out the speed of the dragster.
- Weight balance to help the dragster stay straight.

Keeping these features in mind will help you understand why the dragster is built the way it has been designed. Additionally, the dragster has been designed in a way to allow plenty of customization to be done to it without losing sight of these strategies. The body design has been created to give the look of a dragster, but you could easily modify the design to your own liking, as well as adding to or reducing the weight of the dragster.

In the end, this particular dragster was able to race down a 10-foot-long track in 2.3 seconds. This is awesome because in my other book, *Smart Robotics with LEGO MINDSTORMS Robot Inventor: Learn to Play with the LEGO MINDSTORMS Robot Inventor Kit and Build Creative Robots*, using the LEGO MINDSTORMS Robot Inventor kit 51515 with similar motors and gears, that dragster was only able to achieve 2.7 seconds.

Building the motor frame

The first part of this build is to align the motors with a larger gear to maximize speed:

1. To begin, you will combine two medium motors together using a yellow *3L* yellow axle. In the space between the motors, insert a 36-tooth double-bevel gear:

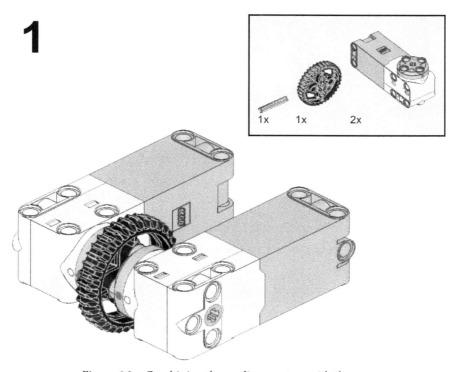

Figure 6.2 – Combining the medium motors with the gear

2. On the side of each of the medium motors, attach two black connector pins on the outside pin holes and connect an azure *13L* beam:

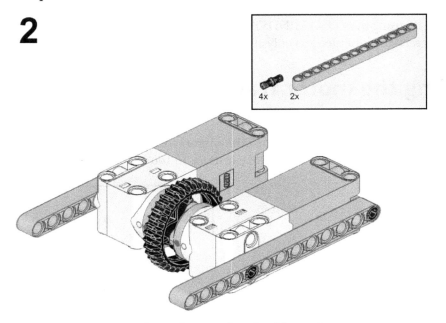

Figure 6.3 – Add a 13L beam to the side of the medium motors

3. You will build a simple frame to house the Intelligent Hub and ensure it does not cause friction with the gear system. The steps to do that are as follows:

A. On the top of each medium motor, add two gray H perpendicular connectors.

B. Connect two black *9L* beams across the motors using the gray H perpendicular connectors.

C. Add two black connector pins to each of the black *9L* beams:

3

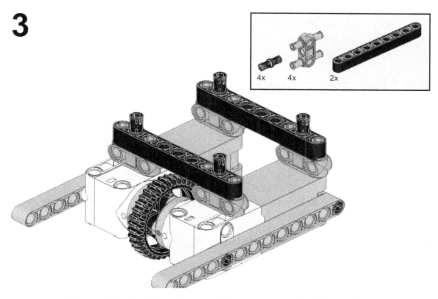

Figure 6.4 – Building the raised frame for the Intelligent Hub

4. Attach the Intelligent Hub using the four black connector pins from the previous step. Secure the Intelligent Hub in place by securing it to the gray H connector pins using azure *3x5L* beams and two black connector pins for each:

4

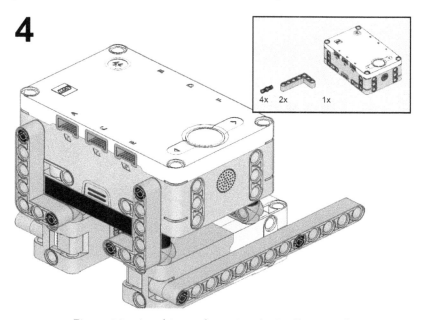

Figure 6.5 – Attaching and securing the Intelligent Hub

That is it for the motor frame. The next step is to begin to add the gears and wheels to increase the speed of your dragster.

Building the gear system and back wheels

Now that the large gear is in place, it is time to add the small gear system to allow the wheels to spin faster:

5.   To begin this part of the build, you will build a sub-model of the gear axle. Follow these steps to do that:

    A. Start with a yellow *7L* axle and add a black 12-tooth double bevel gear into the middle of the axle.

    B. Add two gray bushings on either side of the gear:

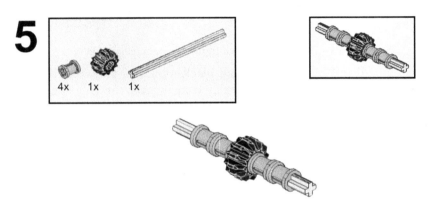

Figure 6.6 – Building out the small gear axle

6. This sub-model should fit between the azure *13L* beams and line up with the larger gear. At this point, nothing is going to hold it into place, but you will take care of that in the next step:

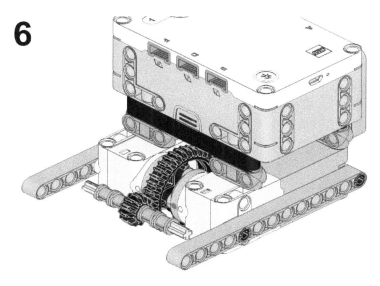

Figure 6.7 – The location of the small gear axle

7. You are going to build another sub-model for the wheels. Follow these steps:

A. Connect two wheels together using a yellow *7L* axle.

B. Add a gray bushing on the side where the yellow *7L* axle is most exposed.

C. On the other side, add a white tooth onto the yellow *7L* axle:

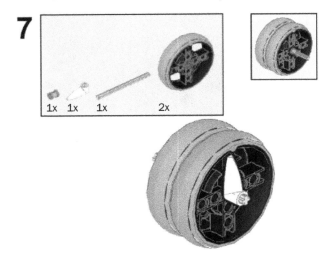

Figure 6.8 – The back wheel sub-model

8.   You will build another one of these so that you have two of these sub-models:

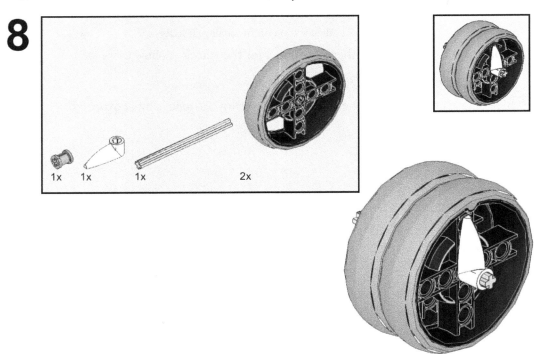

Figure 6.9 – The back wheel sub-model view from the other side

9. You will now connect the back wheels to the small gear axle using the white axle connector. The small 12-tooth gear should align with the large 36-tooth gear:

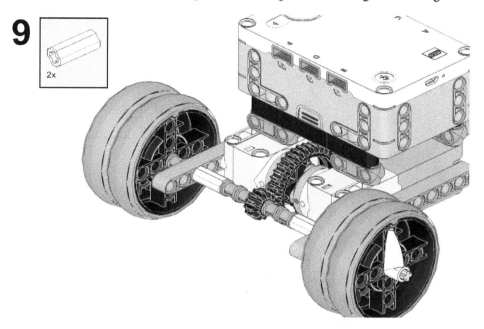

Figure 6.10 – Connecting the wheels to the small gear axle

Now that the gear system is in place, it is time to move toward building out the rest of the dragster body along with the front wheel and sensor.

Building the body of the dragster

To begin the build of the rest of the body of the dragster, you will work on the opposite end to where the gears are located:

10. Add an azure *7L* beam across the medium motors using two black connector pins:

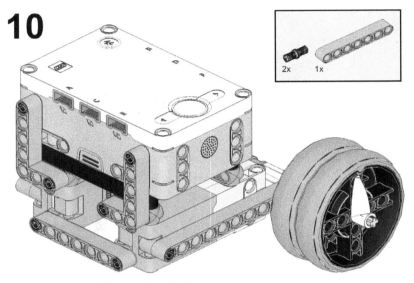

Figure 6.11 – Building brace across motors

11. Connect a purple *7x11* open frame to the azure *7L* beam you installed in the preceding step using two black connector pins:

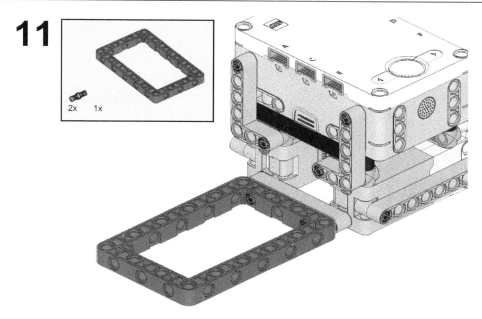

Figure 6.12 – Adding an open frame to the build

12. Add another purple *7x11* open frame to the dragster body using another two black connector pins:

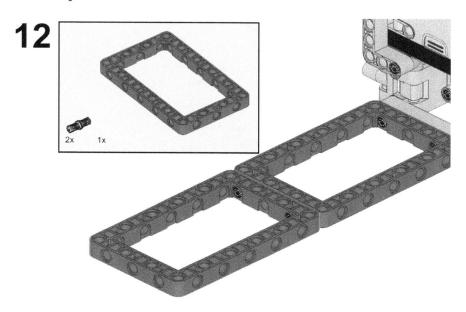

Figure 6.13 – Adding a second open frame to the build

13. Connect an azure *7L* beam to the end of the second purple *7x11* open frame using two black connector pins:

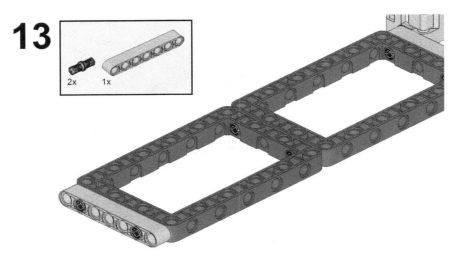

Figure 6.14 – Adding a beam across the open frame

14. Rotate your dragster so that you can see underneath the robot and locate where the purple *7x11* open frame connects to the main frame of the robot. Secure these together using two black T beams, connecting each one using three black connector pins:

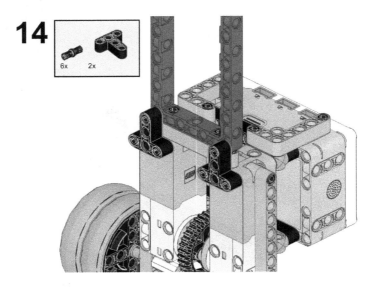

Figure 6.15 – Securing the open frames to the body of the dragster

Now that the frame is built out, you need to create the front wheel sub-model. Your dragster needs a front wheel and, in this build, it will be hidden by the frame of the vehicle:

15. To begin, add two black connector pins to a black *9L* beam, leaving one pin hole open at the end. Do this for two black *9L* beams:

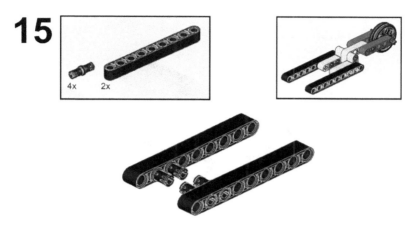

Figure 6.16 – Adding black connector pins to a black 9L beam

16. Attach a yellow *2x4L* beam to each of the black *9L* beams. Connect these two black beams together using a blue connector pin to join the yellow *2x4L beams*, as shown in *Figure 6.17*:

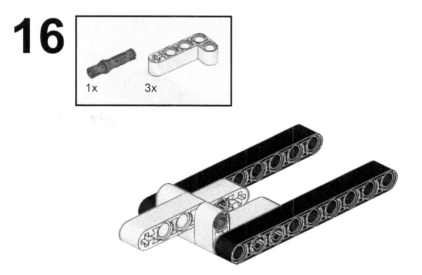

Figure 6.17 – Join the parts together using yellow 2x4L beams

17. At the end of the yellow *2x4L* beam that is exposed, insert a blue connector pin and black pin and axle connector:

Figure 6.18 – Adding connector pins to the end of the yellow 2x4L beam

18. Attach a purple *5L* beam to each side of the yellow *2x4L* beam using the connector pins:

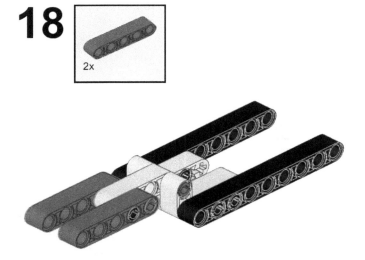

Figure 6.19 – Adding purple 5L beams

19. If you have not already done this with previous builds or creations, take the time to add the rubber tires to the pulley wheel hub from the wheels in your kit:

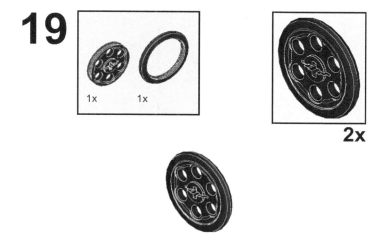

Figure 6.20 – Adding rubber tires to the hub

20. Align the tires together where the axle holes are in the same direction. If you are having issues, then check to ensure the pin holes are lined up properly:

Figure 6.21 – Aligning the wheels

21. You are going to need a black *4L* axle to hold the tires together. The following figure shows the axle pushed through, but you don't need to do this now. Just be sure you can get the axle through the wheels properly:

Figure 6.22 – Testing that the 4L axle fits through the wheels

22. Thread the tires through the purple *5L* beam on the front wheel sub-model and hold it in place using the *4L* axle:

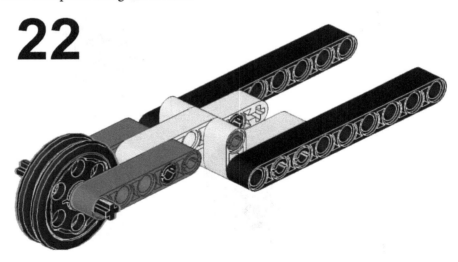

Figure 6.23 – Adding wheels to the sub-model build

23. Your front wheel sub-model is now complete. This sub-model will fit into the purple *7x11* open frame on the dragster:

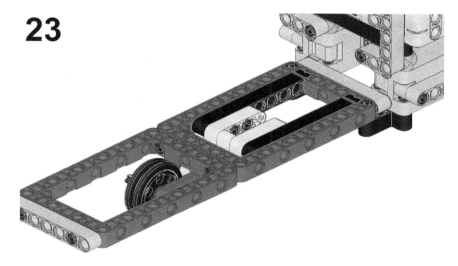

Figure 6.24 – Placing the front wheel sub-model on the open frame

24. Using four gray connector pins with bushes, join the front wheel sub-model to the dragster frame. You will use these pins later, starting at *step 40*, to complete the build of the body frame:

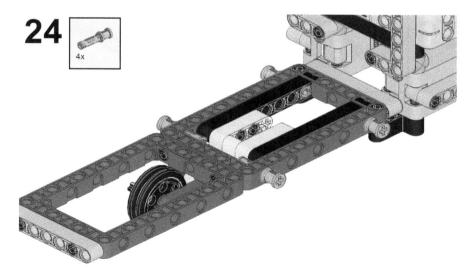

Figure 6.25 – Securing the front wheel sub-model to the dragster frame

25. Turning your focus to the front of the dragster, add four black connector pins to the underside of the purple *7x11* open frame:

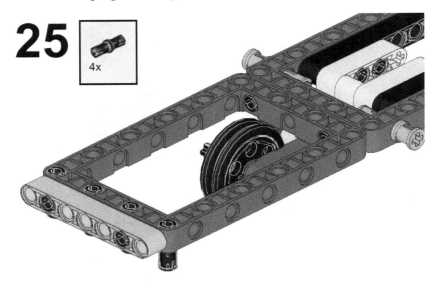

Figure 6.26 – Adding four connector pins to the open frame

26. Add two purple biscuit elements to the four black connector pins. On the inside pin hole of each biscuit element, insert a tan axle pin connector:

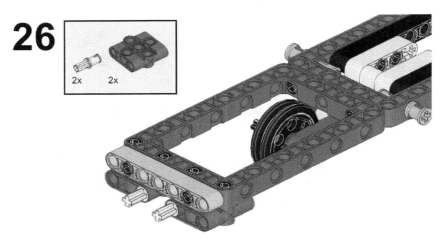

Figure 6.27 – Adding biscuit elements

27. Connect a yellow axle and pin connector *#6* to each of the tan axle pin connectors:

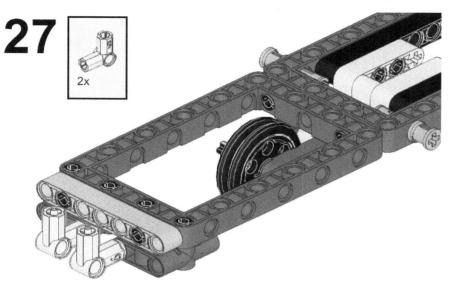

Figure 6.28 – Adding yellow pin and axle connectors #6

28. It is time to hold these elements in place as the tan connector pins are frictionless. The steps are as follows:

A. Add a red axle connector pin to each yellow axle and pin connector #6.

B. Attach a yellow axle and pin connector #6 to the red pins facing outward.

C. Hold these elements in place using a yellow *3L* axle with a bush between them:

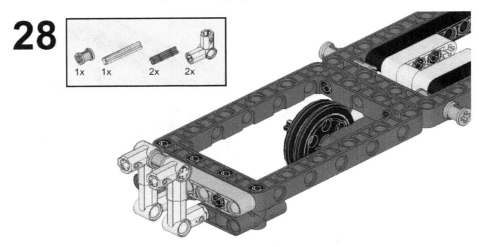

Figure 6.29 – Locking the elements in place

29. Using two more tan pin and axle connector pieces, secure a color sensor to the end of the dragster:

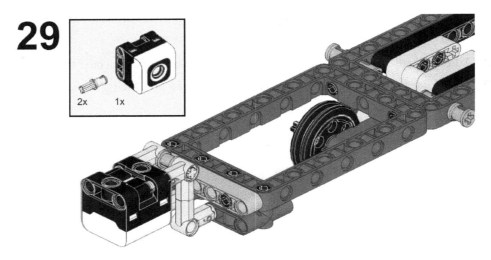

Figure 6.30 – Adding the color sensor

The essentials of the dragster are now complete. You have everything built to allow the dragster to race. Let's add some design elements to spruce it up and make it look a bit better.

Adding the body design

The first part of the body design is to create a space for one of the minifigs in the LEGO SPIKE Prime kit to sit and drive (or choose your own favorite minifig). The steps are as follows:

30. Begin with a black *5x7* open frame and secure it to the purple *7x11* open frame closest to the Intelligent Hub, using four black connector pins:

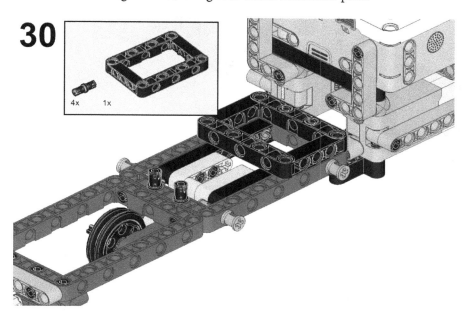

Figure 6.31 – Attaching the black 5x7 open frame to the dragster frame

31. Add another black *5x7* open frame to the same purple *7x11* open frame but secure it to the black open frame you just installed in the previous step, using two blue connector pins and a purple *5L* beam between them:

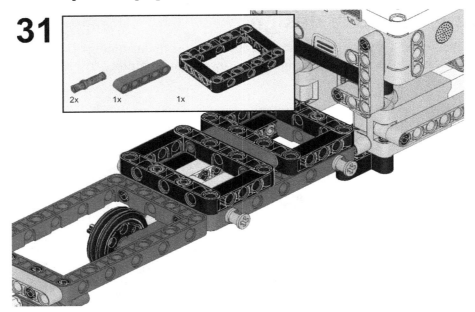

Figure 6.32 – Adding the second black 5x7 open frame to the dragster frame

32. Create four of these elements by combining a gray perpendicular connector pin and a yellow *3L* beam:

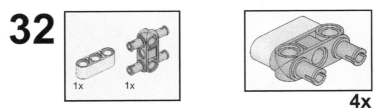

Figure 6.33 – Make four of these elements

33. Secure these four elements into the insides of the black *5x7* open frames. Additionally, add four black connector pins to the top of the gray connector pins inside of the black *5x7* open frame:

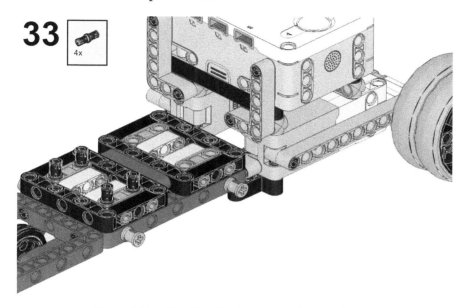

Figure 6.34 – Attaching the elements to the open frames

34. Combine a purple *5L* beam to the azure *8x6x2* curved windscreen using two black connector pins. This will end up being the seat for the minifig:

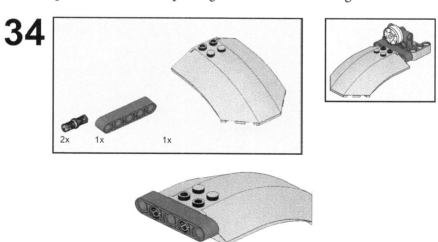

Figure 6.35 – Assembling the minifig seat

35. Connect a gray H frame to a purple biscuit element using a blue connector pin. Attach this to the purple *5L* beam, as shown in *Figure 6.36*:

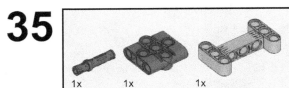

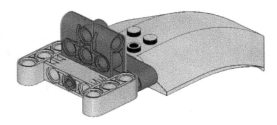

Figure 6.36 – Attaching the H beam to the minifig seat

36. Insert a tan pin and axle connector to the top pin hole of the purple biscuit element. Add a white wheel hub to this pin to serve as a steering wheel for the dragster:

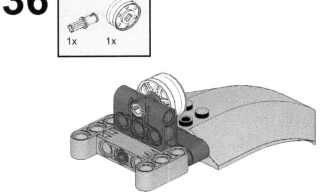

Figure 6.37 – Building the steering wheel

37. This entire steering wheel and minifig seat sub-model will then be attached to the black open frames on the dragster:

37

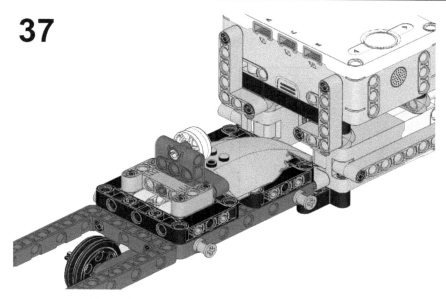

Figure 6.38 – Attaching the driver seat to the dragster

38. Put your focus toward the exposed purple *7x11* open frame. This will be the final phase in building out the dragster:

38

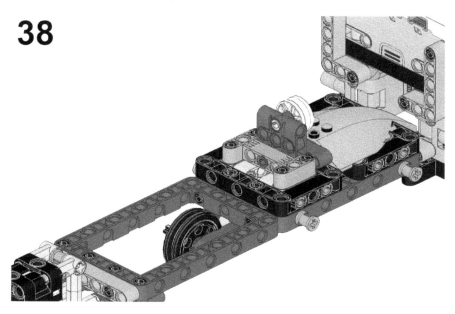

Figure 6.39 – Time to design the front of the dragster

39. Add two azure blue curved panels to either side of the purple *7x11* open frame. Hold each of these curved panels in place using two black connector pins:

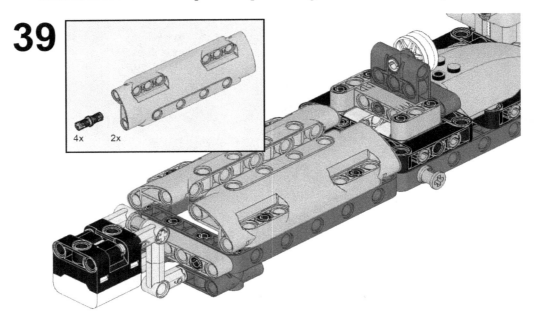

Figure 6.40 – Covering up the wheel with curved panels

40. Before you finish the front of the dragster, you need to add a black *3x11x1* panel plate to the sides of the dragster to create a cohesive body flow for the dragster. Attach the black *3x11x1* panel plate to each side of the dragster using a tan axle pin connector and black axle and pin connector, as shown in *Figure 6.41*:

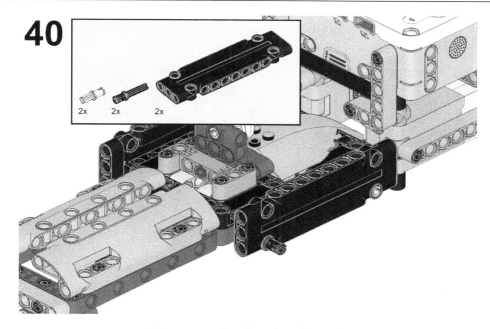

Figure 6.41 – Installing the side panels

41. To build out a solid connection, add two azure *7L* beams to either side of the purple *7x11* open frame using two blue connector pins for each:

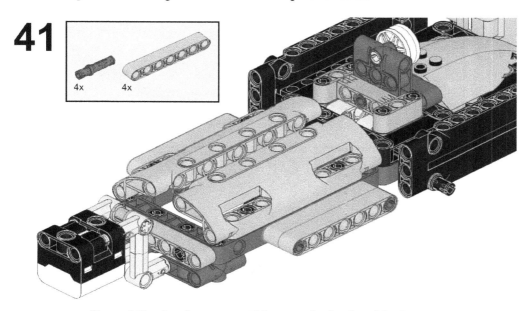

Figure 6.42 – Attach two azure 7L beams to both sides of the dragster

42. Insert a black connector pin to the middle pin hole of the azure *7L* beam on both sides of the dragster:

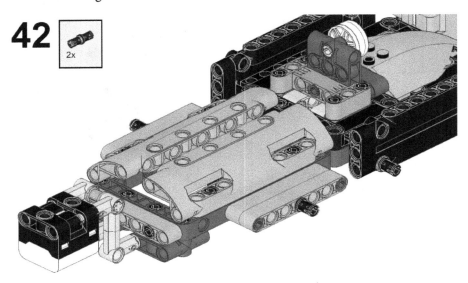

Figure 6.43 – Insert black connector pins

43. Hold everything in place using a black *15L* beam. At the end of the black *15L* beams, click on an azure *8x3x2* wedge to give a finished look to the dragster:

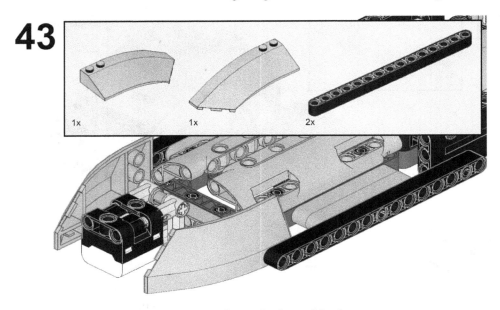

Figure 6.44 – Completing the front of the dragster

44. Lastly, you can add a spoiler to the back of the dragster. This is a very simple one, but keep in mind that you can create and expand this to your liking:

A. Add a tan axle pin connector to the corner pin holes of the Intelligent Hub.

B. Attach a white axle and connector #4 to the tan pins.

C. Attach a red axle connector to the end of the white axle and connector #4.

D. Attach a gray axle and connector #1 to the red axle pin.

E. Connect these elements using a yellow 7L axle and two bushings to hold it in place:

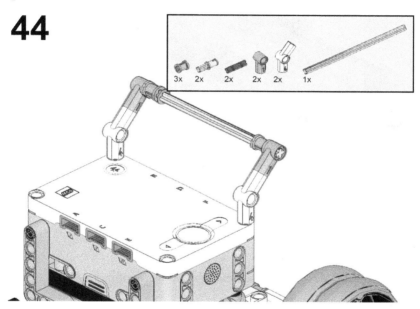

Figure 6.45 – Installing the spoiler

Your dragster should now look like *Figure 6.46*:

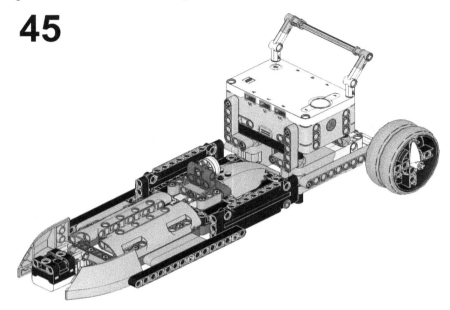

Figure 6.46 – Final view of the dragster

Congratulations! You now have a dragster ready for the track! It is time to program it for action.

Coding the dragster

The code for this project is designed to achieve the following objectives:

- The dragster needs to count down *3-2-1* to make sure people have time to move out of the way.

- The dragster needs to stop when it sees the red (or the color of choice) finish line.

- The dragster needs to display the time to avoid human error when using a stopwatch.

Connecting to the proper ports

Before we begin to code, you need to make sure you have all the parts properly connected to the Intelligent Hub. Start with the motors. The motors should be plugged into ports A and E. It does not matter which motor goes into which port if they are A and E, since the dragster will only move forward and backward.

Your color sensor will go into port C. This sensor will help your dragster identify and stop at the finish line:

Figure 6.47 – The port view in the software

For this program, you will have two code stacks. Additionally, you will be introduced to the broadcast coding blocks to trigger events.

Code stack 1 – countdown timer and broadcast

A dragster race begins with a countdown. You will create a 3-second countdown timer that will trigger the dragster to take off down the track. The steps are as follows:

1. Begin with the yellow events block named **when program starts**.

2. Add a pink movement block named **set movement motors to** and set the motors to ports A and E.

3. Add a purple light block named **turn on for 2 seconds**. Change the graphic to the numeral **3** and change the time to **1** second.

4. Add a purple light block named **set Center Button light to** and change to the color red.

5. Add a light-purple sound block named **start sound** and choose the **Car Idle** sound effect from the library.

6. Add a purple light block named **turn on for 2 seconds**. Change the graphic to the numeral **2** and change the time to **1** second.

7. Add a purple light block named **set Center Button light to** and change to the color yellow.

8. Add a light-purple sound block named **start sound** and choose the **Car Idle** sound effect from the library.

9. Add a purple light block named **turn on for 2 seconds**. Change the graphic to the numeral **1** and change the time to **1** second.

10. Add a purple light block named **set Center Button light to** and change to the color green.

11. Add a light-purple sound block named **start sound** and choose the **Car Vroom** sound effect from the library.

12. Add a yellow event block named **broadcast**. Name the broadcast **start**. This block is used to allow better flow of the code, as well as providing space to expand your code down the road when you experiment and tinker with new ideas.

Here is how this code stack should look:

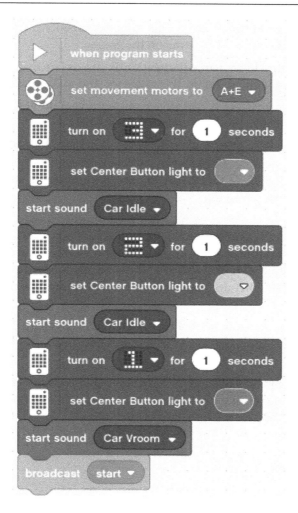

Figure 6.48 – Code stack one

Code stack 2 – activating motors and stopping on the finish line

Now that the first section is done, ending with a broadcast signal of *start*, you need to write code for what happens when the signal is received:

1. Start by creating a new stack using the yellow event block named **when I receive**. Adjust this block to receive a **start** broadcast.

2. Add a blue sensor **reset timer** block. This will reset the internal timer to 0 to get an accurate reading of your dragster time on the course.

3. Add an orange control block named **repeat until**. In this block, drag over a blue sensor block named **is color**. Adjust this to see red and make sure the sensor port is C.

4. Within this **repeat until** block, add a pink **more movement** block named **start moving straight at 50% power.** Adjust this to -100% power.

5. Next, you have to write code for what happens once the color sensor sees red. As soon as the color sensor sees red, the repeat block ceases action and the code moves to the next coding block. In this case, you need the dragster to stop, using the pink movement named **stop moving**.

6. Next, you want to see the time for how fast your dragster moved down the track. Create a variable in the orange variable coding section named **time**. Add a **variable** block named **set variable to 0**. Change the variable to **time**, and for **0**, add a blue sensor block named **timer**. This will grab the time as soon as the dragster crosses the finish line. If you don't do this, then the timer will continue to add time to the overall value.

7. Add an orange control block named **repeat 10** and change it to **3**. In this example, the time will display three times, so the designer has time to walk down the track and read the time being displayed.

8. Within the repeat block, add a purple light block named **write Hello** and add your orange *time* variable block to the **Hello** section. This will display the **timer** variable on the *5x5* LED screen on the Intelligent Hub.

Here is what this code stack looks like:

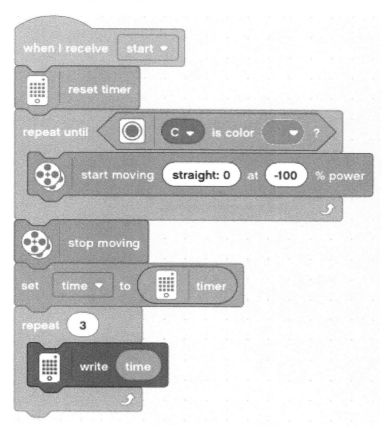

Figure 6.49 – Code stack two

In the end, the entire code will look like this:

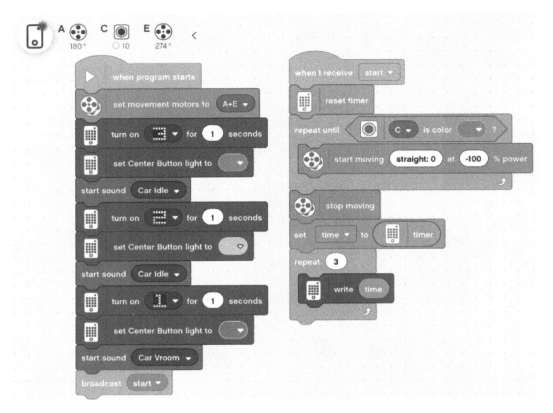

Figure 6.50 – The final code view

Go ahead and give it a try!

You have done it! You should have a working dragster that is full of potential and speed, with plenty of room to modify and tweak to your design needs.

Create a test racetrack. Find a smooth surface to race your dragster. Use a tape measure and measure out a 10-foot track. Using tape, mark the starting line with one color (any color but red) and the finish line with red tape. If you do not have red tape, then use another color and adjust your code accordingly with the color sensor. Enjoy!

Making it your own

And now it is time to hand the dragster over to you. This is where you can take what you have built so far and customize the dragster to your liking. Plenty of pieces have been left in the bin, such as designing aesthetic features, new approaches to making your dragster more intuitive, and plenty of coding options to take the dragster to the next level.

Here are a few ideas to consider:

- What can you change on the body of the dragster to make it look like one you would drive?

- Search online for different dragster designs and tweak this model to another body design.

- Design a better seat for the minifigs to drive the dragster.

- What are other approaches to help your dragster become faster? Less weight, better gearing, and drag resistance reduction are a few things to consider.

Summary

In this chapter, you explored the concept of gearing up to increase the speed of your dragster. You also explored the timer block in the coding as well as the use of the broadcast block. Again, this is one of my favorite build challenges because you can spend hours designing the perfect dragster to become as fast as possible. It is a timeless activity that everyone loves.

In the next chapter, you will be building a game to really challenge the mind.

7
Building a Simon Says Game

Handheld games have long been a pastime favorite. Whether you grew up with basic analog games or started with the glory of digital gaming to the now ever-so-powerful phone apps, you cannot argue how fun games are to the human mind.

In this chapter, you will be building and coding a classic handheld game of Simon Says. This chapter is different from the previous chapters as the focus is less on the actual build and more space is given to the code. The goal of this chapter is to increase your coding skills so that you can go and create a favorite game of your own.

Here is a diagram of what your Simon Says game will look like by the end of this chapter:

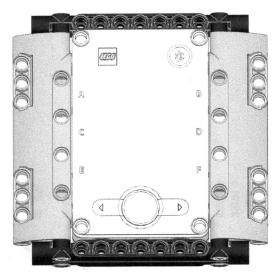

Figure 7.1 – Simon Says handheld game

In this chapter, you will build and program in the following sections:

- Building the handheld game
- Coding Simon Says

Technical requirements

For the building of Simon Says, all you will need is the **SPIKE Prime Kit**. For programming, you will need the **LEGO SPIKE** application/software.

Access to the code can be found here: `https://github.com/PacktPublishing/Design-Innovative-Robots-with-LEGO-SPIKE-Prime/blob/main/Ch.%207%20SIMON%20SAYS%20.llsp`.

You can find the code in action video for this chapter here: `https://bit.ly/2ZhBcw3`

Building the handheld game

This is a very simple build. Please keep in mind that with a game such as Simon Says, there are countless ways to build the device. The build is kept simple so that you can modify it to your own liking:

1. Begin the build by starting with the Intelligent Hub.

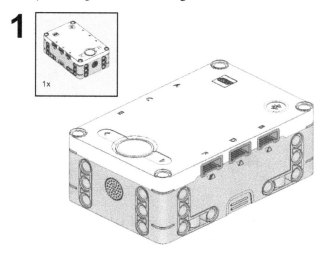

Figure 7.2 – Beginning with the Intelligent Hub

2. On the bottom row of pin holes on the sides of the Intelligent Hub, add two gray H perpendicular connector pins to each side.

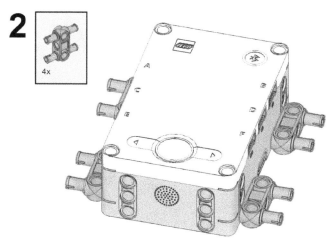

Figure 7.3 – Attaching gray perpendicular connector pins

3. Attach a purple *11L* beam to both sides across the two gray H perpendicular connector pins.

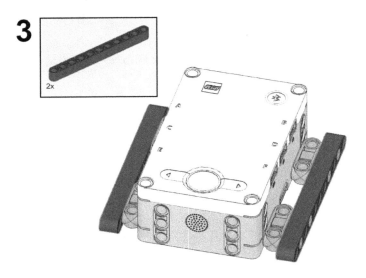

Figure 7.4 – Attaching purple 11L beams

4. Next, add four black connector pins to the top-corner pin holes of the Intelligent Hub.

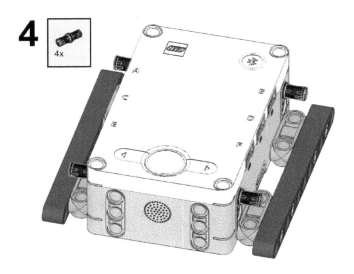

Figure 7.5 – Inserting four black connector pins

5. On either side of the Intelligent Hub, attach an azure curved panel to the black connector pins. On the ends of these azure curved panels, insert black connector pins.

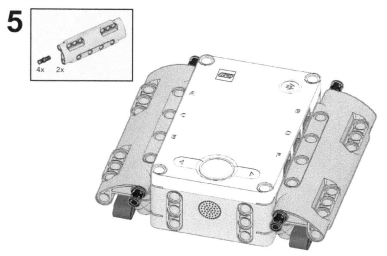

Figure 7.6 – Attaching curved panels to the sides of the Intelligent Hub

6. Secure everything into place using black *3x11* flat panels on either end.

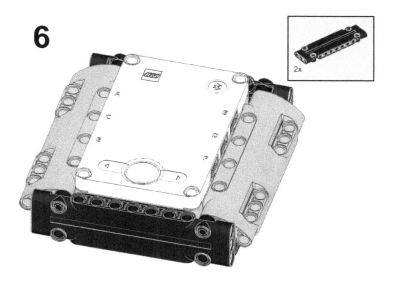

Figure 7.7 – Securing everything in place using black panels

And that is it! A simple handheld gaming device that should now look something like the following figure:

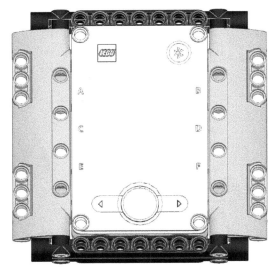

Figure 7.8 – Final build view

Now comes the fun and challenging part of coding the game to be operational for play.

Coding Simon Says

You are going to code a game of Simon Says. The code will be broken down into six parts, as mentioned here:

- Code stack 1 – main code
- Code stack 2 – displaying images for player direction
- Code stack 3 – displaying images and sound for game sequence
- Code stack 4 – player input of answers
- Code stack 5 – game over sequence
- Code stack 6 – correct choice music sequence

All of the code is dependent on the gyro sensor built within the Intelligent Hub. Instead of using colors and external sensors, everything has been designed to use just the Intelligent Hub.

Let's begin!

Code stack 1 – main code

Please keep in mind that while you put this code stack together, some blocks may not work or make sense until you code out the rest of the code stacks. To ease the flow for you, I have decided to code by stacks.

The first part of this stack will clear and reset all the variables to 0. It will also display a smiley face to let the player know the game is activated with the word **Ready?**. This will indicate the sequences are about to begin. The steps to complete this goal are as follows:

1. Begin with the default yellow **Events** block named **when program starts**.

2. Create a variable by clicking on **Make a Variable** in the orange **Variables** block section and name it `score`. Add an orange **Variables** block named **set score to 0**.

3. Create another variable by clicking on **Make a Variable** in the orange **Variables** block section and name it `direction show`. Add an orange **Variables** block named **set score to 0** and change the variable name to `direction show`.

4. Add a purple **Light** block named **M1 turn on smiley face**.

5. Add an orange **Control** block named **wait 1 seconds** and change it to **2 seconds**.

6. Create a variable by clicking on **Make a Variable** in the orange **Variables** block section and name it `charge`. Add an orange **Variables** block named **set charge to 1**.

7. Add a purple **Light** block named **M1 write** and change the word to **Ready?**.

8. Add an orange **Control** block named **wait 1 seconds** and change it to **2 seconds**.

9. Create a list by clicking on **Make a List** in the orange **Variables** block section and name it `colors to remember`. Add a dark-orange **list** block to **Variables** named **delete all of colors to remember**.

When you have completed all of these steps, the beginning of your code stack will look as in *Figure 7.9*:

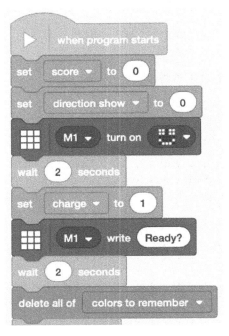

Figure 7.9 – Setting up the game

The next set of instructions will set the parameter for the game to choose random code sequences to try and copy as a player:

1. Under the first section of blocks in this stack, you will now add a yellow **Events** block named **forever**. Within this forever loop, you will program a series of decisions to activate and allow the game to work. Let's begin with the first set of instructions for the loop sequence. Create a variable by clicking on **Make a Variable** in the orange **Variables** block section and name it count. Add an orange **Variables** block named **set count to 0** and change **0** to **1**.

2. Create another variable by clicking on **Make a Variable** in the orange **Variables** block section and name it color selected. Add an orange **Variables** block named **set color selected to 0**, change the setting to a green **Operators** block named **pick random 1 to 10**, and change **10** to **4**.

3. Add a dark-orange **list** block to the orange **Variables** block section named **add thing to answers** and change it to **add color selected to colors to remember**.

4. Add an orange **Control** block named **repeat 10** and change the **10** to a dark-orange **list** block named **length of colors to remember**. After that, do the following:

    A. Add a yellow **Events** block named **broadcast and wait** and add a dark-orange **list** block named **item of colors to remember**. For this block, make sure you add the **count** variable to the item you are pulling from.

    B. Add an orange **Variables** block named **change count by 1**.

5. Below the repeat block sequence, add an orange **Variables** block named **set count to 1**.

6. Add a purple **Light** block named **M1 write Go**. When you have completed all of these steps, the middle section of your code stack will look as in *Figure 7.10*:

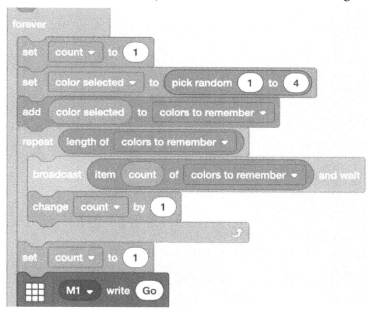

Figure 7.10 – Game choosing sequence of choices

In this final section of this code stack, you will now write the code that will check the inputs from the player against the random sequence created by the program and either advance the player to the next stage if correct or end the game for being wrong:

1. Add an orange **Control** block named **repeat 10** and change the **10** to a dark-orange **list** block named **length of colors to remember**:

    A. Add an orange **Variables** block named **set direction show to 1**.

    B. Add an orange **Variables** block named **set color selected to 0**.

C. Add an orange **Control** block named **wait until**. Add a green **Operators** block of **greater than**. On the left side of the greater than sign, add the variable block named **color selected** and for the right side, make it the numeral **0**.

D. Add an orange **Control** block named **if**. Set the conditions to be a green **Operators** block named **not**. Add another green **Operators** block inside with an equals sign. On the left of the equals sign, insert the **color selected** variable block. On the right of the equals sign, add the **item count of colors to remember** list block:

    i.   Add an orange **Variables** block of **set score to** and insert the **length of colors to remember** list block.

    ii.   Add a yellow **Events** block named **broadcast**. Create a new broadcast message of **game over**.

E. Add an orange **Variables** block named **set direction show to 0**

F. Add an orange **Variables** block named **change count by 1**.

The last two blocks are nested within the forever block.

2.   Create a new variable named `charge`.

3.   Add an orange **Variables** block named **set charge to 1**.

4.   Add a purple **Light** block named **M1 turn on smiley face for 2 seconds**.

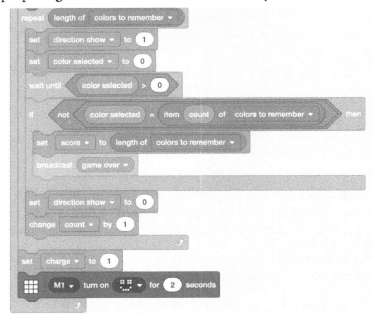

Figure 7.11 – Checking and making decisions based on player input

That completes the first code stack. At this point, the code won't work because you still have to write more code for the variables and then they will be activated.

In all, this is how your first code stack should look:

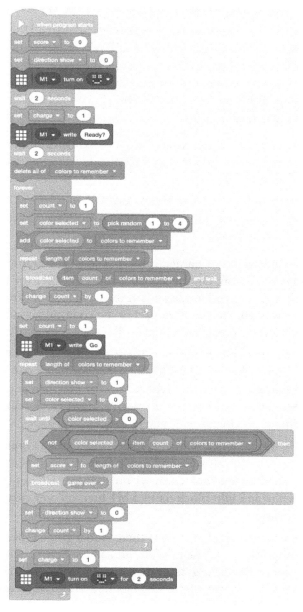

Figure 7.12 – Code stack 1

It is now time to write our second code stack, which will display the angle the player is tilting the Intelligent Hub in order to select the proper sequence.

Code stack 2 – displaying images for player direction

This code stack is for when the **direction show** variable is activated. The code is using a technique called flags where the code is turning on and off certain code stacks throughout the main program. In this case, when playing the game, the player won't be able to access the code sequence if the direction of the tilt of the Intelligent Hub is always showing. Therefore, the code will turn this on when the player is in control and deactivate it when it would cause a glitch in the game:

1. Add a yellow **Events** block named **when**. Add a green **Operators** block of **equal comparison**. On the left side, add the **direction show** variable block and on the right side, add the numeral **1**.

2. Add an orange **Control** block named **repeat until**. Add a green **Operators** block of equal comparison. On the left side, add the **direction show** variable block and on the right side, add the numeral **0**:

    A. Add an orange **Control** block named **if then**. Add a green **Operators** block of greater than comparison. On the left side, add a blue **Sensor** block named **pitch angle**. Change it to **roll angle** and on the right side of the greater than comparison, make it **10**. For the **then** part of the code, add a purple **Light** block named **M1 turn on** and create an image for the right-side activation.

    B. Add an orange **Control** block named **if then**. Add a green **Operators** block of less than comparison. On the left side, add a blue **Sensor** block named **pitch angle**. Change it to **roll angle** and on the right side of the less than comparison, make it **-10**. For the **then** part of the code, add a purple **Light** block named **M1 turn on** and create an image for the left-side activation.

    C. Add an orange **Control** block named **if then**. Add a green **Operators** block of less than comparison. On the left side, add a blue **Sensor** block named **pitch angle**. On the right side of the less than comparison, make it **-10**. For the **then** part of the code, add a purple **Light** block named **M1 turn on** and create an image for the top-side activation.

    D. Add an orange **Control** block named **if then**. Add a green **Operators** block of greater than comparison. On the left side, add a blue **Sensor** block named **pitch angle**. On the right side of the greater than comparison, make it **10**. For the **then** part of the code, add a purple **Light** block named **M1 turn on** and create an image for the bottom-side activation.

The following figure shows what this code stack looks like all put together:

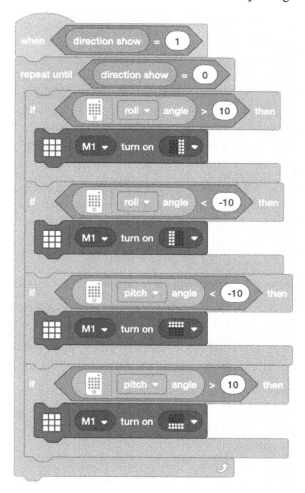

Figure 7.13 – Code stack 2

It is time for our next code stack, which will display the images for the player to remember and input.

Code stack 3 – displaying images and sound for game sequence

This code stack will be a series of four stacks that look the same but have subtle differences depending on what number is chosen randomly. In the main stack, there is a section of code at the beginning where a random number is chosen between 1 and 4. This number gets added to a list, but it also gets broadcasted out. When the number is broadcasted, then the corresponding code stack will activate. In this case, it will display the image to remember, and a music note to correspond with the position. Additionally, there is a step where the screen goes blank so that when the same direction is chosen twice in a row, it helps visually to know this has happened.

In this code example, there are four positions for the game – up, down, right, and left. You will be taken through how to code these stacks but remember that you can right-click the first stack, duplicate it, and just change the parameters, saving time of dragging over each block. Then, you simply duplicate the other three, making the changes to your liking. The following steps show you how to code each of the images for directions:

1. Add a yellow **Events** block called **when I receive** and choose the numeral **1**.

2. Add a purple **Light** block named **M1 turn on** and create an image for the right-side activation.

3. Add a light-purple **Sound** block named **play beep 60 for 0.2 seconds**.

4. Add an orange **Control** block named **wait 1 seconds** and change it to **.5** seconds.

5. Add a purple **Light** block named **M1 turn on** and turn off all the LEDs.

6. Add an orange **Control** block named **wait 1 seconds** and change it to **.5** seconds. You should have code that looks like this:

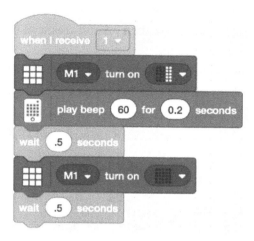

Figure 7.14 – Right-side activation

7. Add a new yellow **Events** block called **when I receive** and choose the numeral **2**.

8. Add a purple **Light** block named **M1 turn on** and create an image for the left-side activation.

9. Add a light-purple **Sound** block called **play beep 62 for 0.2 seconds**.

10. Add orange **Control** block named **wait 1 seconds** and change it to **.5** seconds.

11. Add a purple **Light** block named **M1 turn on** and turn off all LEDs.

12. Add an orange **Control** block named **wait 1 seconds** and change it to **.5** seconds.

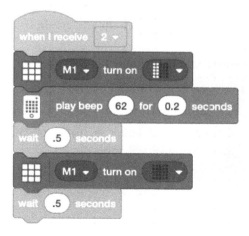

Figure 7.15 – Left-side activation

13. Add a yellow **Events** block called **when I receive** and choose the numeral **3**.

14. Add a purple **Light** block named **M1 turn on** and create an image for the top-side activation.

15. Add a light-purple **Sound** block called **play beep 64 for 0.2 seconds**.

16. Add orange **Control** block named **wait 1 seconds** and change it to **.5** seconds.

17. Add a purple **Light** block named **M1 turn on** and turn off all LEDs.

18. Add an orange **Control** block named **wait 1 seconds** and change it to **.5** seconds.

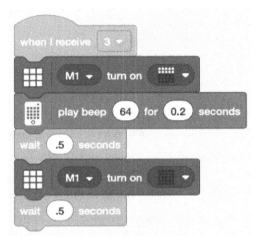

Figure 7.16 – Up activation

19. Add a yellow **Events** block called **when I receive** and choose the numeral **4**.

20. Add a purple **Light** block named **M1 turn on** and create an image for the down activation.

21. Add a light-purple **Sound** block called **play beep 65 for 0.2 seconds**.

22. Add an orange **Control** block named **wait 1 seconds** and change it to **.5** seconds.

23. Add a purple **Light** block named **M1 turn on** and turn off all LEDs.

24. Add an orange **Control** block named **wait 1 seconds** and change it to **.5** seconds. You should have something that looks like this:

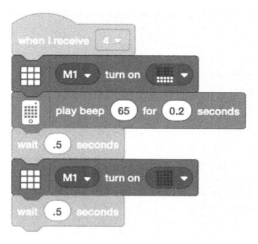

Figure 7.17 – Down activation

In all, you should have four of these stacks. They will look as in the following figure:

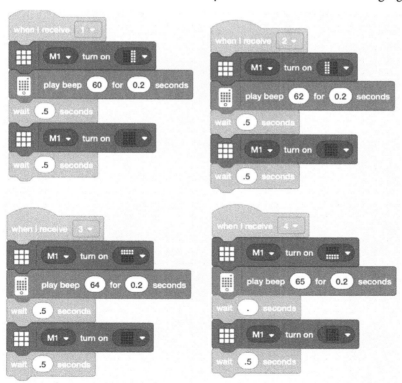

Figure 7.18 – Code stack 3

The next section of code stacks will be designed to allow the user to input their sequence in order.

Code stack 4 – player input of answers

This section of code stacks is similar to what we just did in code stack 2 and code stack 3. What needs to happen is that when it is time for the player to input the sequence, there needs to be a way to activate the guess and score it against the game sequence. To do this, the code will use the right button on the Intelligent Hub combined with the angle the player is tilting the Intelligent Hub. This helps the player be sure their guess is what they want.

Again, like the code stacks in part 3, once you create one, you can right-click and duplicate and just change the parameters, saving the time of dragging over each block:

1.   Add a yellow **Events** block named **when**. Add a green **Operators** block called **and**:

   A.   On the left side of the **and** operator block, add a green greater than comparison **Operators** block:

      i.   To the left side of the comparison, add a blue **Sensors** block named **pitch angle**. Change it to **roll angle**.

      ii.   To the right side of the comparison, add the numeral **10**.

   B.   On the right side of the **and** operator block, add a blue **Sensor** block named **is left button pressed**. Change this block to **is right button pressed**.

2.   Add a purple **Sound** block named **play beep 60 for 0.2 seconds**.

3.   Add an orange **Variables** block named **set color selected to 1**.

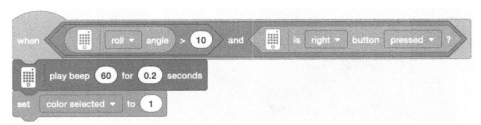

Figure 7.19 – Player input for the right direction

4. Add a yellow **Events** block named **when**. Add a green **Operators** block of **and**. Next, do the following:

A. On the left side of the **and** operator block, add a green **Operators** block for less than comparison:

 i. To the left side of the comparison, add a blue **Sensors** block named **pitch angle**. Change it to **roll angle**.

 ii. To the right side of the comparison, add the numeral **-10**.

B. On the right side of the **and** operator block, add a blue **Sensor** block named **is left button pressed**. Change this block to **is right button pressed**.

5. Add a purple **Sound** block named **play beep 62 for 0.2 seconds**.

6. Add an orange **Variables** block named **set color selected to 2**. You should have something that looks like this:

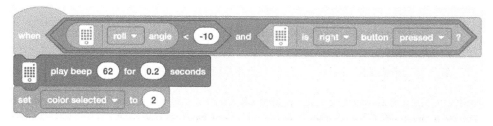

Figure 7.20 – Player input for the left direction

7. Add a yellow **Events** block named **when**. Add a green **Operators** block for **and**. Next, do the following:

A. On the left side of the **and** operator block, add a green less than comparison **Operators** block. Following this, do the following:

 i. To the left side of the comparison, add a blue **Sensors** block named **pitch angle**.

 ii. To the right side of the comparison, add the numeral **-10**.

B. On the right side of the **and** operator block, add a blue **Sensor** block named **is left button pressed**. Change this block to **is right button pressed**.

8. Add a purple **Sound** block named **play beep 64 for 0.2 seconds**.

9. Add an orange **Variables** block named **set color selected to 3**. You should have something that looks like this:

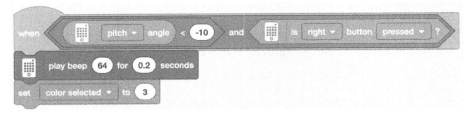

Figure 7.21 – Player input for the up direction

10. Add a yellow **Events** block named **when**. Next, add a green **Operators** block for **and**. Following this, do the following:

   A. On the left side of the **and** operator block, add a green **Operators** block for greater than comparison:

   i.   To the left side of the comparison, add a blue **Sensors** block named **pitch angle**.

   ii. To the right side of the comparison, add the numeral **10**.

   B. On the right side of the **and** operator block, add a blue **Sensor** block named **is left button pressed**. Change this block to **is right button pressed**.

11. Add a purple **Sound** block named **play beep 65 for 0.2 seconds**.

12. Add an orange **Variables** block named **set color selected to 4**.

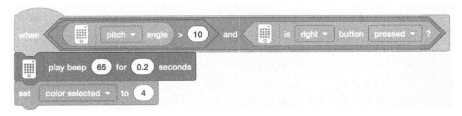

Figure 7.22 – Player input for the down direction

Here are the four code stacks of this section all complete:

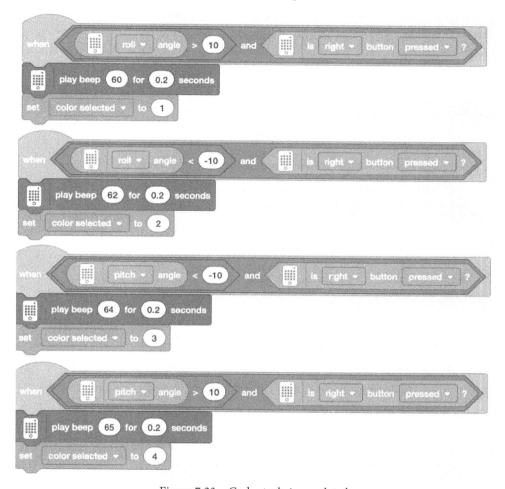

Figure 7.23 – Code stack 4 completed

You have completed the code stacks for this section. Your game is now playable and interactive. There are still two stacks left to complete, but the interaction components have been completed.

It is time to add the final essential code stack for this program to work. This is the programming for when the gameplay is over.

Code stack 5 – game over sequence

In the main code, there is a section that checks the player input against the sequence generated by the code. When a player inputs the wrong sequence item, then there is a broadcast block that triggers the game over sequence. This code stack is where you will code what happens when the game is over.

In this example, there is a music sequence programmed that you might recognize along with the posting of the score of the player.

The following instructions will provide you with a sample jingle for the game to play:

1. Add a yellow **Events** block named **when I receive game over**.
2. Add an orange **Control** block named **stop other stacks**.
3. Add a green **Music** block named **set tempo to 60** and change it to **110**.
4. Add a green **Music** block named **set instrument to piano** and change it to **Organ**.
5. Add a green **Music** block named **play note 60 for .25 beats** and change it to **note 64**.
6. Add a green **Music** block named **play note 60 for .25 beats**.
7. Add a green **Music** block named **rest for .25 beats**.
8. Add a green **Music** block named **play note 60 for .25 beats** and change it to **note 57**.
9. Add a green **Music** block named **play note 60 for .25 beats** and change it to **note 55**.
10. Add a green **Music** block named **rest for .25 beats** and change it to **.75**.
11. Add a green **Music** block named **play note 60 for .25 beats** and change it to **note 64**.
12. Add an orange **Control** block named **repeat 10** and change it to **2**:

    A. Add a green **Music** block named **play note 60 for .25 beats** and change it to **note 64**.

    B. Add a green **Music** block named **rest for .25 beats**.

13. Add a green **Music** block named **play note 60 for .25 beats**.
14. Add a green **Music** block named **play note 60 for .25 beats** and change it to **note 64**.
15. Add a green **Music** block named **rest for .25 beats**.
16. Add a green **Music** block named **play note 60 for .25 beats** and change it to **note 67**.
17. Add a green **Music** block named **rest for .25 beats** and change it to **.75**.
18. Add a green **Music** block named **play note 60 for .25 beats** and change it to **note 55**.
19. Add a green **Music** block named **rest for .25 beats** and change it to **.75**.
20. Add a purple **Light** block named **M1 write** and input the word **Score**.

21. Add a purple **Light** block named **M1 write** and input an orange **score** variable block.

22. Add an orange **Control** block named **wait 1 seconds** and change it to **3**.

23. Add an orange **Control** block named **stop all**.

Here is the completed code stack for game over:

Figure 7.24 – Code stack 5 completed

You are now down to one last code stack. The final code stack is a music sequence used to start the game and when the player inputs the sequence correctly.

Code stack 6 – correct choice music sequence

This code stack is for a fun little jingle that plays when the game starts as well as when the player inputs the correct sequence for each level.

The following instructions will provide you with a sample song jingle for the game to play:

1. Add a yellow **Events** block named **when**. Add a green equal comparison **Operators** block. On the left side, add a **charge** variable block and on the right side, the numeral **1**.

2. Add a green **Music** block named **set tempo to 60** and change it to **70**.

3. Add a green **Music** block named **set instrument to piano** and change it to **Organ**.

4. Add a green **Music** block named **play note 60 for .25 beats** and change it to **note 50 for .15 beats**.

5. Add a green **Music** block named **rest for .01 beats**.

6. Add a green **Music** block named **play note 60 for .25 beats** and change it to **note 53 for .15 beats**.

7. Add a green **Music** block named **rest for .01 beats**.

8. Add a green **Music** block named **play note 60 for .25 beats** and change it to **note 57 for .15 beats**.

9. Add a green **Music** block named **rest for .01 beats**.

10. Add a green **Music** block named **play note 60 for .25 beats** and change it to **note 60 for .3 beats**.

11. Add a green **Music** block named **rest for .2 beats**.

12. Add a green **Music** block named **play note 60 for .25 beats** and change it to **note 57 for .2 beats**.

13. Add a green **Music** block named **rest for .01 beats**.

14. Add a green **Music** block named **play note 60 for .25 beats** and change it to **.7 beats**.

15. Add an orange **Variables** block named **set charge to 0**.

Here is the completed code stack for charge:

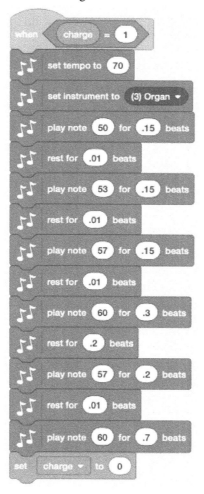

Figure 7.25 – Code stack 6 completed

Whew! You did it. Now it is time to play the game and have fun. As you play the game, think about all the new ways you could modify this game to something new and even better than this version!

Making it your own

Here is the best part of every chapter and build. What will you do to remix the build and the code to make it unique to your own creative power? While Simon Says is a standard game, there are a lot of ways to enhance your build.

Here are a couple of ideas to consider applying to this game:

- What new graphics could you display besides the solid bars? Arrows? Images?
- What new sound effects could you add? What are your favorite jingles from movies, cartoons, or video games that you could add?
- Could you change this from a Simon Says game to a Bop It game? Instead of tilting the Intelligent Hub, could you have the player activate one of the gesture controls such as shaken, tapped, or falling?
- What if you used a touch sensor or even rotated a motor to add new levels of difficulty to the game?

Summary

This chapter spent a great deal of time on coding using broadcast blocks, variables, lists, and music. The goal was to open your eyes to some new ways to think about robots and coding. As you continue to play Simon Says, what will you do next? How can you combine all the learning from this chapter as well as the previous chapters to create something engaging, fun, and so awesome that everyone will want to learn from you?

Remember, all of these builds in the book are designed to inspire you. Take what you have learned and make something even better and more unique. When you design one of these builds, or remix it to your own liking, please share it with the hashtag `#legospikecreations`. The world can't wait to see your genius!

Packt.com

Subscribe to our online digital library for full access to over 7,000 books and videos, as well as industry leading tools to help you plan your personal development and advance your career. For more information, please visit our website.

Why subscribe?

- Spend less time learning and more time coding with practical eBooks and Videos from over 4,000 industry professionals

- Improve your learning with Skill Plans built especially for you

- Get a free eBook or video every month

- Fully searchable for easy access to vital information

- Copy and paste, print, and bookmark content

Did you know that Packt offers eBook versions of every book published, with PDF and ePub files available? You can upgrade to the eBook version at packt.com and as a print book customer, you are entitled to a discount on the eBook copy. Get in touch with us at customercare@packtpub.com for more details.

At www.packt.com, you can also read a collection of free technical articles, sign up for a range of free newsletters, and receive exclusive discounts and offers on Packt books and eBooks.

Other Books You May Enjoy

If you enjoyed this book, you may be interested in these other books by Packt:

Smart Robotics with LEGO MINDSTORMS Robot Inventor

Aaron Maurer

ISBN: 9781800568402

- Discover how the Robot Inventor kit works, and explore its parts and the elements inside them
- Delve into the block coding language used to build robots
- Find out how to create interactive robots with the help of sensors
- Understand the importance of real-world robots in today's landscape
- Recognize different ways to build new ideas based on existing solutions
- Design basic to advanced level robots using the Robot Inventor kit

Build and Code Creative Robots with LEGO BOOST

Ashwin Shah

ISBN: 9781801075572

- Unbox the LEGO BOOST kit and understand how to get started
- Build simple robots with gears and sensors
- Discover the right parts to assemble your robots
- Program your BOOST robot using the Scratch 3.0 programming language
- Understand complex mechanisms for advanced robots
- Develop engaging and intelligent robots using electronic and non-electronic components
- Create more than 10 complete robotics projects from scratch
- Develop logical thinking and unleash your creativity

Packt is searching for authors like you

If you're interested in becoming an author for Packt, please visit authors.
packtpub.com and apply today. We have worked with thousands of developers and
tech professionals, just like you, to help them share their insight with the global tech
community. You can make a general application, apply for a specific hot topic that we are
recruiting an author for, or submit your own idea.

Share your thoughts

Now you've finished *Design Innovative Robots with LEGO SPIKE Prime*, we'd love to hear
your thoughts! If you purchased the book from Amazon, please click here to go straight
to the Amazon review page for this book and share your feedback or leave a review on the
site that you purchased it from.

Your review is important to us and the tech community and will help us make sure we're
delivering excellent quality content.

Index